Plant Life

Plant Life

Roland Ennos and Elizabeth Sheffield

School of Biological Sciences
3.614 Stopford Building
University of Manchester
Manchester M13 9PT
UK

Website: www.biomed.man.ac.uk
Email : roland.ennos@man.ac.uk
*   liz.sheffield@man.ac.uk*

Blackwell
Science

© 2000 by
Blackwell Science Ltd
Editorial Offices:
Osney Mead, Oxford OX2 0EL
25 John Street, London WC1N 2BS
23 Ainslie Place, Edinburgh EH3 6AJ
350 Main Street, Malden
  MA 02148 5018, USA
54 University Street, Carlton
  Victoria 3053, Australia
10, rue Casimir Delavigne
  75006 Paris, France

Other Editorial Offices:
Blackwell Wissenschafts-Verlag GmbH
Kurfürstendamm 57
10707 Berlin, Germany

Blackwell Science KK
MG Kodenmacho Building
7–10 Kodenmacho Nihombashi
Chuo-ku, Tokyo 104, Japan

The right of the Authors to be
identified as the Authors of this Work
has been asserted in accordance
with the Copyright, Designs and
Patents Act 1988.

First published 2000

Set by Graphicraft Ltd, Hong Kong
Printed and bound in Great Britain
at the University Press, Cambridge

For further information on
Blackwell Science, visit our website:
www.blackwell-science.com

DISTRIBUTORS

Marston Book Services Ltd
PO Box 269
Abingdon, Oxon OX14 4YN
(*Orders*: Tel: 01235 465500
         Fax: 01235 465555)

USA
Blackwell Science, Inc.
Commerce Place
350 Main Street
Malden, MA 02148 5018
(*Orders*: Tel: 800 759 6102
         781 388 8250
         Fax: 781 388 8255)

Canada
Login Brothers Book Company
324 Saulteaux Crescent
Winnipeg, Manitoba R3J 3T2
(*Orders*: Tel: 204 837 2987)

Australia
Blackwell Science Pty Ltd
54 University Street
Carlton, Victoria 3053
(*Orders*: Tel: 3 9347 0300
         Fax: 3 9347 5001)

A catalogue record for this title
is available from the British Library

ISBN 0-86542-737-2 ✔

Library of Congress
Cataloging-in-publication Data

Ennos, A.R.
   Plant life / Roland Ennos and Elizabeth Sheffield
     p. cm.
   ISBN 0-86542-737-2
   1. Botany. 2. Plants—Evolution. I. Sheffield, Elizabeth.
   II. Title.

   QK47.E56 2000
   580—dc21   00-023721

The Blackwell Science logo is a
trade mark of Blackwell Science Ltd,
registered at the United Kingdom
Trade Marks Registry

Contents

Contents

Colour plate section falls between pp. 110 and 111.

Preface

Photosynthetic organisms are vital and fascinating. They are the factories which trap the energy which drives almost all ecosystems. But plants are also extremely diverse in both their size and their shape, ranging in form from tiny cyanobacteria a matter of micrometres in diameter to 100 metre high trees. Despite this we felt that they always seemed to draw the short straw in biology courses. In many school curricula the concentration is almost entirely on the process of photosynthesis, and the delightful variety of the forms and lifestyles of plants is swept under the carpet. At the other extreme, plant diversity courses at universities tend to plough remorselessly through all the different taxonomic groups of plants. They emphasize a bewildering variety of morphological and reproductive characteristics, but fail to treat plants as living organisms. Consequently courses tend to become exercises in rote learning rather than in understanding plant life.

We felt that it was high time to redress the balance by writing a book with a strong evolutionary theme, and fortunately in many ways the timing is ideal for such an enterprise. Recent advances in cladistics and molecular systematics have given us a far better idea of how plants are related to each other and how they evolved. At the same time popular books, and television series such as *The Private Life of Plants* have shown that treating plants as animals are treated —as competing and struggling organisms—pays dividends. They come to life and even seem to exhibit individual personalities. This book takes this approach, and, using some of the excellent zoology texts as a model, investigates plants from an evolutionary perspective to tell what we feel is the fascinating story of plant life.

Of course one cannot understand the pattern of plant evolution without knowing something about how evolution works or how it can be studied. The first section of the book therefore outlines evolutionary and phylogenetic theory. This theory is then used to enlighten the story, told in the second section of the book, of how the major taxonomic groups of plants live and how and why they evolved. But plants from different taxonomic groups are by no means evenly distributed around the globe. The final section examines the varied vegetation types found in the different parts of the globe, and investigates how and why plant diversity varies. It seeks to explain why certain groups of plants are more successful and examines the many ingenious means all sorts of plants use to survive.

The book has taken too long to produce! We would like to thank the succession of commissioning editors from Blackwell Science, Simon Rallison, Susan Sternberg and Ian Sherman, for their interest and encouragement throughout the process. We would also like to thank the reviewers John Dodge, Karl Niklas, Thomas Speck and Barry Thomas for their helpful comments and suggestions on early drafts of the manuscript. Any remaining errors are ours. Much of the merit of a book of this sort must be due to the quality of the illustrative material. In our case we have been enormously fortunate to have been supplied with this by our knowledgeable and helpful colleagues: Fred Rumsey, who produced the excellent set of line drawings; Sean Edwards, who supplied most of the photographs; Elizabeth Cutter, Sally Huxham and Bill Chaloner. We thank all of them for supplying us with illustrations, some of which we did not even know we wanted! Finally we would like to thank our families: Yvonne, Russ and Max for all their encouragement and forbearance throughout the long gestation period of this book.

Roland Ennos and Liz Sheffield

Part 1

—————— CHAPTER 1 ——————

Evolution and plant diversity

1.1 INTRODUCTION—THE DIVERSITY OF PLANT LIFE

There are two overriding impressions you receive if you survey the world of plants. The first is one of bewildering diversity, despite the fact that most plants produce food by the same mechanism—photosynthesis. This is true most obviously in numerical terms, because there are probably over half a million species of photosynthetic organisms. But there is also a vast diversity of forms, since photosynthetic organisms range from tiny unicellular algae under a micrometre in diameter to huge multicellular trees over 100 m tall and 100 tonnes in weight.

The second impression, which you get when you look more closely at each species, is the perfection of its adaptations to its own particular lifestyle. Each photosynthetic organism seems to be well suited, and apparently well designed, to live in its own particular habitat. Cacti, for instance, are adapted to the dry desert habitat by having thick, barrel-like stems which are well suited for storing water; and spines, which help protect the water from being stolen by animals. They make a striking contrast with kelps which have adapted to the subtidal environment in which mechanical forces dominate. Their flexible fronds are ideally suited to resist the battering of waves and they are securely cemented to the sea bed by a holdfast.

It is the aim of this book not only to *catalogue* the diversity of photosynthetic organisms and *describe* their adaptations, but also to explain *how* this has been achieved. The key to understanding both diversity and adaptation is Darwin's theory of **evolution by natural selection**, and therefore this book will have a strong evolutionary slant. Modern plants are not just a random collection of organisms which may

be conveniently placed by taxonomists into arbitrary groups, but are the results of millions of years of struggle for life. As we shall see, this has resulted not only in the evolution of large numbers of species, but has also driven evolution in directions which are, with the benefit of hindsight, readily explicable. Once plants invaded land, for instance, selection for taller and taller plants would have made the evolution of trees almost inevitable.

1.2 FACTORS AFFECTING PLANT EVOLUTION

1.2.1 The physical environment

There is no doubt that the physical environment in which a plant grows strongly constrains its evolution. All plants require water, nutrients and light, and to obtain them their body form has to change with the habitat.

Because marine organisms are surrounded by water they do not have the problem of desiccation and have no need to develop special water-conducting tissues. However, because water rapidly attenuates light and because dissolved gases diffuse only slowly, they need adaptations to obtain adequate light and nutrients. Consequently many marine forms are unicellular and motile, using flagella to stay up in the upper illuminated regions and to move to nutrient-rich areas.

In contrast, although the air that surrounds land plants rapidly supplies the gases they require, it allows them to desiccate rapidly. Most land plants have therefore had to develop special root systems for absorbing water from the soil, and vascular systems to conduct it to their leaves. As a consequence they are by necessity immobile.

Some habitats are also so extreme that only a few particularly well-adapted organisms can survive in them. For example, encrusting algae are the only forms that can survive on severely wave-swept shores. Similarly rocks in deserts are colonized only by the few species of lichens that can tolerate the near-permanent desiccation.

1.2.2 The limitations of physical explanations

Adaptation to the physical environment can explain much, but by no means all, about the diversity of plants. First, there are far more species of plants than there are physical habitats. Second, many plants are not found in parts of the world where they are perfectly capable of growing. Cacti, for instance, will also thrive in much wetter places than the deserts in which they are found and they are even absent from the deserts of the Old World. Finally, and most crucially, many of the most important attributes of plants cannot be explained by physical factors alone. There is no physical reason, for instance, why trees should have trunks and branches; because they do not photosynthesize, they will be merely a drain on a tree's resources. If a physicist were to design an ideal 'efficient' land plant it would resemble the simple liverwort *Pellia*, being a simple layer of photosynthetic cells lying on the soil surface.

1.2.3 Biological interactions

Because physical factors cannot alone explain the diversity of plant form it is clear that other factors must be involved. It is generally accepted that the evolution of plants has also been greatly driven by their struggle for life with other organisms. The great diversity and adaptive perfection of plants is therefore due more to their biological interactions (Fig. 1.1) rather than just to their adaptations to their physical environment.

Competition

The most obvious of the selection pressures that have driven (and continue to drive) the evolution of plants

is **competition** with other plants. Our ideal low-lying land plant would be rapidly outcompeted by any plants that managed to grow above; they could shade it out by producing multicellular stems to hold up the leaves. As well as competition for light, plant evolution also has been driven by competition for the other essentials of life: water, nutrients and even space itself. Some plants can even discourage growth of competitors near them by secreting growth-inhibiting substances, a process called **allelopathy**.

Defence

Other important selection pressures are driven by attempts by other organisms to exploit the photosynthesis of a plant. Plants have evolved defences which prevent this happening, but there are many ways in which plants can be exploited. Most obviously, plants can be eaten by animals, a process known as **herbivory**. Considering the huge number of herbivores and the inability of most plants to move, it is surprising that any plants survive at all! But plants can also be exploited by smaller organisms which penetrate their tissues and exploit them from the inside. When this invasion is practised by tiny organisms such as viruses, bacteria and fungi this results in **disease.** When it is carried out by small animals, in contrast, it is usually described as **endophagy**, and when another plant carries out the process it is known as **parasitism**. Many of the evolutionary novelties devised by plants are defensive mechanisms which reduce their losses by these processes.

Exploitation

Yet other selection pressures are driven by the attempts of plants themselves to exploit other life forms. These include **parasitism** of other plants; the consumption of animals to obtain energy and nutrients, a process known as **carnivory**; and associations with other organisms to exchange benefits, a process known as **mutualism**. Mutualistic partnerships are widespread and have been extremely important in the evolution of plants. They can occur between two species of plants; between plants and bacteria, as in the root nodules of legumes; between plants and fungi, as in lichens or mycorrhizas; and between

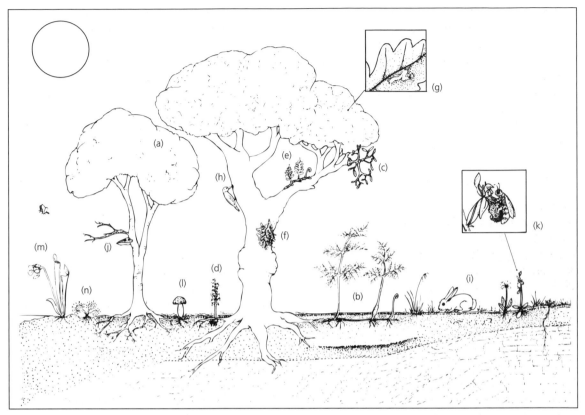

Fig. 1.1 A hypothetical region of woodland in which there are a wide variety of biological interactions between the plants and other organisms. There is **competition** (a) as the smaller tree is being shaded out by the taller one and root competition is occurring between the herbs on the right. The bracken fern (b) to the right of the trees prevents competition by other plants by **allelopathy**, using chemicals to prevent them growing below it. A mistletoe (c) is **parasitizing** the branch of the larger tree, while the **saprophyte** *Neottia* (d) is living on decaying humus produced by the tree. Support is being obtained from the tree by an epiphytic fern (e), and lichen (f), which is itself a **mutualistic** association between a fungus and an alga. Other organisms are also exploiting the trees' photosynthesis. Leaf miners (g) are eating away the larger tree's leaves, while a woodpecker (h) carves out its nest from the tree trunk. The grasses are the victims of **herbivory** by the rabbit (i), while the tree on the left is the victim of a fungal **disease** (j). There are also other mutualistic interactions. The orchid (k) on the right is being **pollinated** by a bee which is attempting to copulate with it, while the smaller tree has a mycorrhizal fungus (l) on its roots. Some plants are even fighting back against animals. The **carnivorous** pitcher plant (m) and sundew (n) on the left have apparatus to capture and digest insects.

plants and animals, as in corals and in the insect pollination of flowers. As we shall see, such mutualistic associations long ago even gave rise to plant cells themselves.

1.2.4 Coevolution

Whatever the type of the interaction between organisms, it usually results in **coevolution** between the

organisms, a form of evolution which is driven by biological rather than physical necessity. For instance, trees have had to grow taller and taller over evolutionary history just to outcompete their neighbours even though the trunk wood itself is unproductive. Similarly, insect-pollinated plants have had to develop larger and more colourful flowers to attract pollinators and increasingly reward them with more nectar. The result of these interactions between organisms, even those that live together in a mutually beneficial relationship, is an **evolutionary arms race**, in which each organism struggles to gain the maximum possible advantage. A meadow is not just a peaceful collection of pretty flowers but a battlefield of warring organisms, each fighting for its life and its chance to pass on its genes to future generations. Botany is the study of the ingenious solutions plants have come up with to survive in this battlefield.

1.3 THE PROCESS OF EVOLUTION

1.3.1 The theory of evolution by natural selection

Before we can make sense of the diversity of plant life it is essential to understand the process of evolution by natural selection, which was first described by Darwin and outlined in his book *On the Origin of Species* (1859). Darwin's theory can be summarized logically in a simple four-stage argument.

1 More individuals are produced than can survive.

2 There is a struggle for existence.

3 Individuals show variation. Those with favourable characteristics are more likely to survive and reproduce.

4 Because selected varieties produce similar offspring to themselves, these varieties will become more abundant.

Therefore, given competition, variation and inheritance, evolutionary change over time is *inevitable*. It was Darwin's genius to recognize the central role of variation, which is often a destructive phenomenon, in producing progressive change. There is no doubt that his ideas were strongly influenced by his knowledge of domestic animals and of the breeding of pigeons. In these animals, new breeds can be produced by selecting for breeding only those individuals which possess the desired traits. In a similar way, Darwin argued, nature would select organisms with better characteristics for survival, by killing off their competitors. His use of the term **natural selection** was chosen deliberately to contrast with the **artificial selection** imposed by breeders on their stock.

1.3.2 Inheritance, genes and microevolution

Darwin had one major problem: his ignorance of a convincing mechanism for inheritance. The findings of Mendel, which were rediscovered only after Darwin's death, remedied this defect and showed that the inherited material was particulate in nature: something which is essential if natural selection is to work. Mendel found that the form and behaviour (or **phenotype**) of an organism is controlled by large numbers of particles called **genes** which are passed, apart from a few mutations, unaltered through the generations. Together, these genes make up an organism's **genotype**.

Discrete characters

Studies of the small-scale **microevolution** have shown that natural selection can alter a population of organisms in two ways. The first involves **discrete characters** which are influenced by a single gene: characters like the colour and texture of Mendel's peas or the tolerance of certain grasses to metal pollution. If one of two possible forms is favoured by the environment the frequency of the favoured genes will increase. Eventually it may even sweep through the entire population to **fixation**. For instance when grasses were grown on copper-mine waste (Fig. 1.2a), plants with a gene for tolerance to copper grew better and left more descendants. Consequently the frequency of the tolerance gene increased over time.

Continuous characters

The second way in which selection can result in evolution involves **continuous characters**, such as the height, diameter and seed size of plants, which are influenced by a large number of genes, each of which has a small effect. Without selection there will be some variation in these characters because the sum of the effect of the genes will often be higher or lower than the mean, just by chance. Usually the numbers of organisms at particular values will follow a normal distribution. If selection pressure is applied to these characters, the relative numbers of genes in the population will be altered and the population will

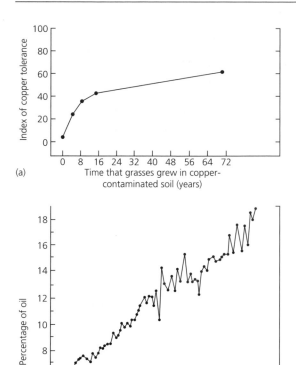

(a)

(b)

Fig. 1.2 Examples of evolution. (a) Response of creeping bent grass *Agrostis stolonifera* to copper contamination. The index of copper tolerance was higher in populations of plants that had been living longer in areas contaminated with copper. This shows that natural selection had resulted in evolution which improved the chance of survival of the population. (Redrawn from Berg, 1997.) (b) Response of maize *Zea mays* to artificial selection for high or low oil content. Over 80 generations the former had almost quadrupled oil content and the latter reduced it by about 90%. (Redrawn from Ridley, 1993.)

be changed. For instance, if herbivores preferentially graze taller grasses because they are easier to eat, the proportion of genes for shortness that are passed to the next generation will be increased and the population will become shorter. A long-term experiment aimed at altering the oil content of maize seeds (Fig. 1.2b) has shown that selection over many

generations can produce extremely large changes very rapidly.

1.3.3 Cytogenetics and sex

Of course, we now know that genes are not just abstract ideas but are real entities which are located in the **chromosomes** which are found in almost all cells. At the molecular scale, following the discoveries of Crick and Watson, we also now know that the genes are coded regions of the molecule **deoxyribonucleic acid (DNA)** which occurs in the famous double helix. However, the ways in which the DNA is packaged within cells differs between the simple **prokaryotes** such as bacteria and the larger, more complex **eukaryotes**, which include unicellular and multicellular algae and all land plants.

Prokaryotes

The prokaryotes have only a single copy of each gene, all of which are encoded onto a single long circular strand of DNA, often referred to as a bacterial **chromosome**, which floats free in the cytoplasm. Additional shorter circular **plasmids** may also be present. Prokaryotes divide asexually by the process of **mitosis** (Fig. 1.3) which produces two nearly identical daughter cells. If the copying mechanism for DNA was perfect, there would be a limit to evolution; selection would reduce the variation within the population and eventually all the surviving organisms would be identical. Instead, rare mistakes in the copying process, which are known as **mutations**, maintain the genetic variation and ensure that evolution can continue. Limited transfer of genetic material is also possible between individuals.

Eukaryotes

In the larger and more slowly reproducing eukaryotes, rare mutations would not be sufficient to ensure rapid enough evolution. One way of increasing the rate of evolution might be to increase the mutation rate of the genes. Unfortunately, though, this has the drawback that, because most mutations are harmful, few of the offspring would survive. Instead, eukaryotes have evolved a much more

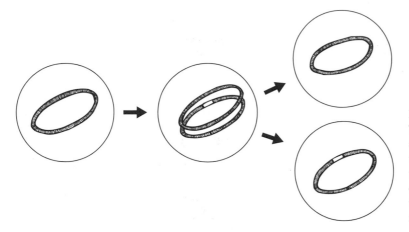

Fig. 1.3 Mitosis in a prokaryote cell. The bacterial chromosome is replicated and the cell then divides, one copy of the chromosome moving into each of the daughter cells. Note that the copying mechanism is not exact, so that one daughter chromosome has a mutation (white area).

sophisticated (and enjoyable!) way of maintaining variability—sex.

The cells of eukaryotic organisms can typically be found in two different states: the cells may be either **haploid,** in which case they contain a single copy of each gene; or **diploid** in which case they contain *two* copies. Each gene is found in a linear chromosome, several of which are usually found within the nucleus, but while haploid cells contain one copy of each chromosome, diploid cells contain two. Both haploid and diploid cells can divide, like bacteria, by mitosis, but mutations are mostly eliminated by special **restriction enzymes** which correct any mistakes that have been made. The variability is produced by allowing genetic information to be transferred between homologous chromosomes when diploid cells divide in a special process called **meiosis** (Fig. 1.4). This process actually involves two separate divisions of the cells. In the first division, each chromosome doubles up, as in mitosis. Then, however, something odd occurs. Instead of splitting, the two pairs of chromosomes line up with each other and may exchange some genetic material. Only then does cell division occur to produce two diploid cells with pairs of mixed chromosomes. Finally these cells divide again to leave four haploid cells each with a single totally original copy of each chromosome.

Meiosis can produce an almost infinite number of new combinations of genes on which natural selection can operate without the drawback of an overly high mutation rate. But meiosis always produces haploid cells. To reform diploid cells which can go through the process again, two haploid cells need to go through another process, **fusion**, in which two cells meet, join together and the nuclei fuse to produce a nucleus with two copies of each chromosome. Sex therefore requires alternate meiosis and fusion.

1.4 LIFE CYCLES OF EUKARYOTES

Apart from its effect on the variability of the offspring an organism produces, the process of sex has important implications for the life cycles of eukaryotic organisms. Because meiosis and fusion must alternate, producing haploid and diploid cells in turn, the life cycles of these organisms are necessarily complex. And the process is made even more complex because both haploid and diploid cells can also divide by mitosis. Haploid cells can either fuse immediately after they have been formed by meiosis, or may divide by mitosis many times first. Similarly diploid cells can undergo meiosis immediately to form new haploid cells or they may divide by mitosis many times first. The different life cycles seen in eukaryotes (Fig. 1.5) are the result simply of different patterns of meiosis, mitosis and fusion.

1.4.1 Zygotic meiosis

If the diploid **zygote** splits by meiosis immediately after it has been formed (Fig. 1.5a), a process known as **zygotic meiosis,** the organism will spend most of

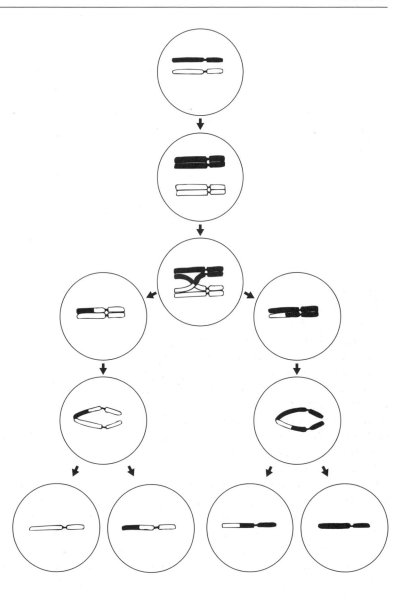

Fig. 1.4 Meiosis in a eukaryote cell. The first stage is a doubling up of each homologous chromosome in the diploid cell. The next involves the lining up of the homologous pairs and exchange of material. The cell then splits to give two diploid cells with pairs of mixed chromosomes. The final cell division produces four haploid cells, each with a single novel copy of the chromosome.

its life in the haploid state. The haploid cells can divide in two different ways which will produce very different sorts of haploid organisms.

The daughter cells of the meiosis can divide and separate from each other to produce generations of single-celled organisms. Eventually some of these organisms may fuse to recreate the **zygote**. Alternatively the daughter cells can remain attached to each other after cell division to create a multicellular organism. Eventually some of the cells of this organism will form gametes which will fuse with gametes from a different organism to re-create the **zygote**. In both cases the zygote is the only diploid cell.

Life cycles involving zygotic meiosis were probably the first to evolve, since the earliest eukaryotes, like prokaryotes, must have been haploid. Zygotic meiosis is found today in algae, such as the unicellular *Chlamydomonas* (Chapter 3), and the multicellular *Chara* (Chapter 4). Fungi also use this method.

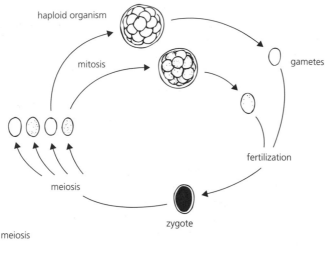

(a) Zygotic meiosis

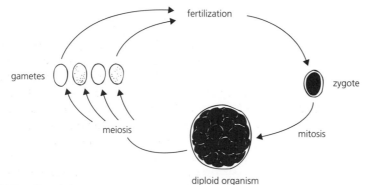

(b) Gametic meiosis

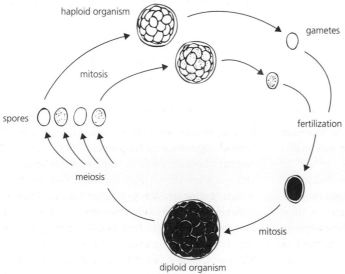

(c) Sporic meiosis

Fig. 1.5 Diagrams of the three types of eukaryote life cycle. In these diagrams the diploid phase is black, while the male and female haploid stages are white and light grey. Some green algae and fungi undergo **zygotic meiosis** (a) in which the zygote divides by meiosis immediately after being formed to give haploid cells. These divide by mitosis to produce multicellular haploid organisms or more single-celled haploid organisms. These eventually produce gametes which fuse to reform the zygote. Some brown algae and animals undergo **gametic meiosis** (b) in which the haploid gametes are formed by meiosis from a multicellular diploid organism. These gametes fuse immediately after they are formed to produce a zygote which divides by mitosis to produce another diploid organism. Some algae and all plants undergo **sporic meiosis** (c) in which both gametes and zygotes divide by mitosis to produce multicellular organisms. The diploid **sporophyte** produces haploid spores by meiosis, and these divide to produce the multicellular **gametophyte** which eventually produces gametes. Because this produces two different generations of adult organisms this sort of life cycle is often known as **alternation of generations**.

1.4.2 Gametic meiosis

If the haploid **gametes** fuse together immediately after they have formed (Fig. 1.5b), the organism will spend most of its life in the diploid state. Again, the form of the diploid organism that forms will depend on the pattern of cell division.

If the diploid cells separate after each cell division, generations of unicellular organisms will be produced. Alternatively if the cells remain attached a multicellular organism will be produced. Eventually, some of the cells will undergo **gametic meiosis** to recreate gametes which fuse with ones from other organisms. In both cases the gametes are the only haploid cells.

Gametic meiosis is characteristic of both multicellular animals such as ourselves and some unicellular protists. It has also developed in some brown algae such as *Fucus* (Chapter 4).

1.4.3 Sporic meiosis

The most complex sort of life cycles occur when both haploid and diploid cells can undergo mitosis (Fig. 1.5c). In this case meiosis of diploid cells produces haploid **spores** rather than gametes. These are released and divide mitotically many times to produce a **haploid organism**. Eventually this produces gametes which fuse to produce a diploid zygote. This in turn divides by mitosis to produce a **diploid organism**. Eventually some of this organism's cells will undergo **sporic meiosis** to recreate the **spores**. In life cycles with this pattern of meiosis therefore there are two **alternating generations** of organisms: one haploid, the other diploid.

Sporic meiosis is characteristic of many algae and all the higher plants. The two generations: the haploid **gametophyte** (so called because it produces gametes) and the diploid **sporophyte** (so called because it produces spores) often look similar in algae, and are then said to be **isomorphic**. However, in many algae and in all plants mutations have occurred which are expressed in only one of the generations. As a result the gametophyte and sporophyte have come to be very different from each other and are said to be **heteromorphic**.

Different generations have become dominant in different groups, as we shall see. In mosses and liverworts (Chapter 5) the gametophyte is dominant, and the sporophyte is usually dependent on it for nutrition. In vascular plants such as ferns (Chapter 6), in contrast, it is the sporophyte that is dominant. This process has been taken to such an extreme in the seed plants, for instance the conifers and angiosperms (Chapter 7), that the gametophytes are microscopic and totally dependent on the sporophyte for nutrition. The females are minute organisms which are held within the ovary of the flower, while the male gametophytes are held within the even tinier pollen grains. As a result, most seed plants seem at first sight to reproduce in a very similar way to animals, but the processes are very different.

1.4.4 Implications of life cycles

The success of organisms depends just as much on their reproductive efficiency as on the factors that affect survival of the existing organisms. The efficiency of reproduction is greatly affected by environmental conditions. It is particularly difficult for unicellular gametes to disperse on dry land, for instance, because desiccation becomes a big problem. As we shall see, therefore, plants' life cycles greatly influence which environments they can effectively colonize. For this reason the evolution of plant life cycles will be discussed throughout this book as we encounter each new group of organisms.

1.5 MECHANISMS OF ADAPTATION

1.5.1 Evolution of complex characters

It is at first glance hard to imagine how the small-scale changes in gene frequency which are caused by natural selection could have produced the vast array of sophisticated adaptations possessed by modern plants. The process must be very slow. It must be remembered, however, that photosynthetic organisms first appeared over 3 billion years ago, and the first eukaryotic organisms over a billion years ago. Natural selection has therefore had plenty of time over which to operate.

Bearing in mind that each stage in evolution has to be an improvement on the previous one, it is also difficult to see how complex organs such as flowers could be produced by the gradual process of evolution by natural selection. Surely half a flower would not be any use to a plant? However, careful thought can allow us to put forward quite plausible ideas about how natural selection could have generated flowers. Many species of insect eat pollen because it is an excellent source of protein. Presumably early flying insects such as beetles would have inadvertently spread pollen from one reproductive structure to another as a result. More insects could also have been attracted by sugar within the pollen drop of early seed plants. Plants that produced this sugar solution would get improved pollen transfer both because it would attract more insects and because each insect would eat less pollen. They would consequently be more likely to pass on their genes.

A change that might improve the efficiency of the pollen transfer process would be the evolution of a signal to the insects that pollen and sugar solution were available. The leaves that surrounded the reproductive parts of the plant could have become more prominent by gradually replacing their chlorophyll with other pigments. The flowers with the more visible leaves would have been visited more frequently by insects and would be more likely to have their pollen transferred and hence to have more offspring. The genes for prominent leaves would therefore become more common, so over time the leaves might become brighter, change shape and so evolve into petals. Similarly, the amount of sugar solution produced by the plant would also gradually increase, because plants with larger amounts of sugar would also be more frequently visited. The end result would be the evolution of sophisticated nectaries. Over time therefore the competition between flowers to attract pollinating insects could produce the beautifully adapted flowers we see today.

1.5.2 Preadaptations

Frequently, large changes in evolution can be brought about when the function of an existing organ is changed. The flower petals, which attract insects, for instance, are probably modified leaves.

Similarly, the tendrils which are used by many climbing plants to grip supports (Fig. 1.6) have evolved from petioles, which were used to bear leaves. Structures that become used in a different way from their original function are known as **preadaptations**. They frequently show extensive evolutionary change after the alteration in function to 'fine tune' them to their new role.

It is often easy to detect when such changes have occurred. One way is to use a comparative approach and examine a range of organisms. Related species may retain the ancestral character. Magnolias, for instance, have flowers that look much more like simple leaf stalks than other flowering plants. Similarly, some climbing plants like *Clematis* climb using tendrils (Fig. 1.6b) which still bear leaves. A second way is to examine mutant forms of the plant, since the original genes may still be present, only masked by later additions. In many mutant peas, for instance, the tendrils can be replaced by fully leafed petioles, showing that the two structures are **homologous**. No structure or organ has yet been found whose evolution cannot be explained by the process of evolution by natural selection. Until one has, therefore, it is safe to assume that evolution by natural selection is the cause of **adaptation**.

1.6 MECHANISMS OF SPECIATION

1.6.1 Reproductive isolation and speciation

A second puzzle for botanists is how so many separate species of plants could have evolved. This is because although natural selection might drive plants of a single species to become different, when they mate with each other, meiosis would constantly be mixing up their genes again and would keep the population uniform. Therefore in a single interbreeding population, although change and adaptation are possible, the evolution of diversity would not be.

The crucial step that allows a species to split up into two or more new ones is for its population to be split up into **reproductively isolated** subpopulations. Once they no longer breed with each other

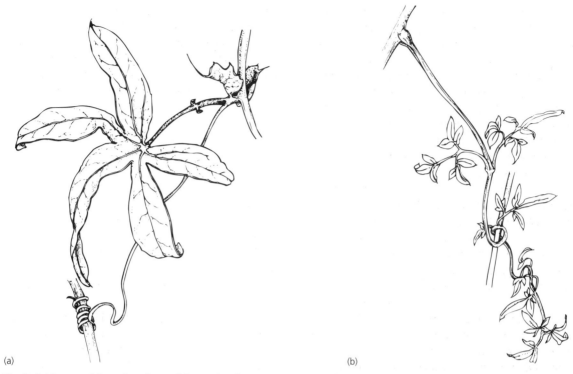

Fig. 1.6 Many tendrils such as those of the passion flower *Passiflora caerulea* (a) are modified petioles. This ancestry is shown up by plants such as *Clematis thibetica* (b) in which the tendrils still bear leaves.

they will be able to alter independently and start to become different. If they breed apart for long enough they may be so different when they meet again that they are unable to breed with each other. Fertilization may be prevented or the hybrids produced between the two forms may be inviable or sterile. If so, **speciation** will have occurred; two or more isolated breeding populations (or **species**) will have been produced where only one existed before.

1.6.2 Allopatric speciation

Clearly the easiest way for populations to become isolated from each other is geographically: populations may be isolated on islands within an ocean; in lakes within a continent; on mountains separated by valleys; in valleys separated by mountains; or in isolated patches of one habitat within another. Much of the speciation of plants probably did take place

as a result of such isolation, particularly in the small populations which often survive at the extreme edges of a species' range. Speciation that occurs as a result of this form of isolation is known as **allopatric speciation**.

There are many examples of plant species that were almost certainly formed in this way. Isolated oceanic islands contain many **endemics**, species found only on that island but similar to forms that grow on the mainland. These species probably formed by allopatric speciation. Often whole groups of closely related plants are found in chains of recently formed volcanic islands, just like Darwin's finches on the Galapagos islands. A good example are the silverswords of Hawaii: originally small daisy-like flowers which must have speciated within the last few million years, as the islands emerged. They have evolved into plants as diverse as large rosette plants, shrubs, lianas, and even trees.

1.6.3 Sympatric speciation

Many similar species often occur in the same place and show no sign of having been isolated from each other geographically. In their case the speciation must have occurred following the reproductive isolation of plants which were growing in the same place, a process called **sympatric speciation**. There are two main ways in which this can occur. In plants which are insect pollinated, speciation can occur as a result of the insects' preferences. If two subpopulations of plants are preferentially visited by two subpopulations or species of insects, they can start to evolve in different ways and can quickly speciate. This mechanism is probably responsible for the evolution of the large number of species of orchids, since these are plants which have particularly sophisticated associations with single species of insect pollinators (see Chapter 7).

The majority of sympatric speciation, however, is thought to occur by a mechanism which is much commoner in plants than in animals: **polyploidy**. This process results in the multiplication of whole sets of chromosomes, and its commonness in plants is probably related to their ability both to self-fertilize and to propagate vegetatively.

Autopolyploidy

Autopolyploidy (Fig. 1.7a) results from straightforward doubling of the chromosome number in a single species of plant. This can happen in two ways. Cells can fail to divide in mitosis, so directly producing sporophyte cells which have four identical chromosomes. Alternatively, cells can fail to divide in meiosis, so producing diploid gametes which contain two identical chromosomes. These can then self-fertilize to form a sporophyte with four identical chromosomes.

Whichever way it is formed, the **tetraploid** plant will be self-fertile. Meiosis is possible because each of the four copies of each chromosome can pair up with one of the three others. The diploid gametes which are formed will then be able to fuse with other diploid gametes to produce another tetraploid zygote. However, the tetraploid will *not* be able to produce fertile offspring from a mating with the

original species because the gametes would fuse to form a **triploid** cell which would not be able to divide properly. Since it forms a new isolated breeding population, the new tetraploid will therefore actually be a new species.

In fact, autopolyploids are rare in nature, possibly because, since each chromosome has three others with which to pair, mistakes in meiosis are common. Most autotetraploids are infertile and rely on asexual reproduction to survive. The best known example is probably the cultivated potato *Solanum tuberosum*, which arose from diploid forms in its native South America. It is nowadays spread by humans in the form of its edible tubers.

Allopolyploids

Allopolyploids, which are far more common in nature, are formed by the **hybridization** (Fig. 1.7b) of two fairly closely related species, followed by a doubling in their chromosome number. This results in a plant which has a genome in which each chromosome has only one similar fellow with which to pair. Allopolyploids are therefore far more frequently self-fertile than autopolyploids. They are also reproductively isolated from both of their parent species for the same reason as autotetraploids, since fusion of the gametes would produce a triploid plant in just the same way. The result of allopolyploidy is therefore a new and very fertile species.

There is no doubt that allopolyploidy is a very important factor in the speciation of plants, and it has been estimated that around 60% of all plant species are polyploids. Since polyploidy usually involves hybridization between species it is therefore somewhat inappropriate to think of the evolution of plants as being like a branching tree. Since many branches join up again it is more like a branching net.

Allopolyploidy can also be a very fast process as it can occur within a matter of a few generations. Many cases of recent speciation by allopolyploidy have been authenticated, including the evolution of the highly successful salt-marsh grass *Spartina anglica*. This species was formed near Southampton following the hybridization in the second half of the nineteenth century of two species, the native British *S. maritima* and the introduced American species

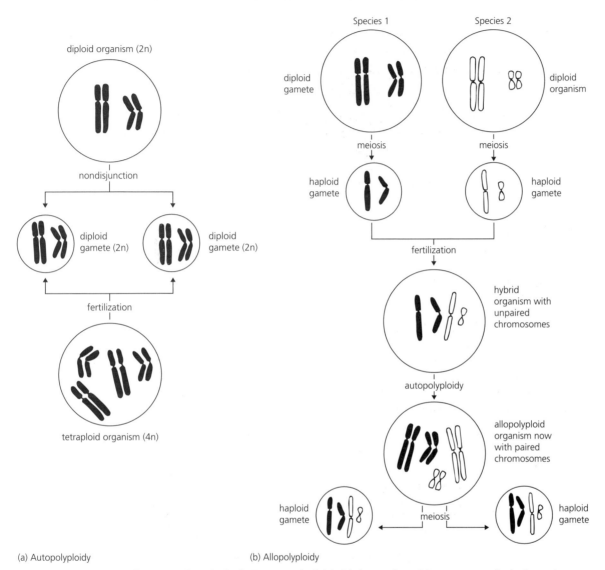

(a) Autopolyploidy   (b) Allopolyploidy

Fig. 1.7 Two methods of speciation by polyploidy. In **autopolyploidy** (a) the number of chromosomes of a single species can be doubled if the chromosomes do not separate during meiosis (non-disjunction). This produces diploid gametes which produce a tetraploid individual with four copies of each chromosome. This individual will be capable of sexual reproduction with others like it, but cannot breed with members of the original diploid parent species. In **allopolyploidy** (b) a new fertile species can be produced in two stages. Hybridization between two different species can produce a hybrid which is sterile because the chromosomes cannot pair in meiosis. If this hybrid later undergoes autopolyploidy, doubling chromosome number, the chromosomes can pair in meiosis and a new fertile species results.

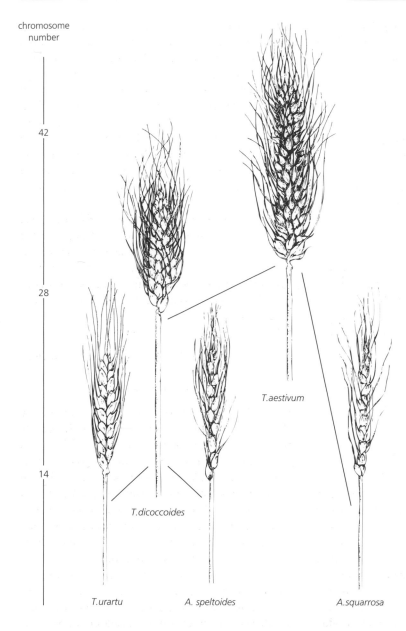

chromosome
number

42

28

14

T.aestivum

T.dicoccoides

T.urartu   *A. speltoides*

A.squarrosa

Fig. 1.8 The evolution of wheat by allopolyploidy. Tetraploid emmer wheat *Triticum dicoccoides* was first formed in the Fertile Crescent following a cross between two plants, each with 14 chromosomes: wild einkhorn *T. urartu* and a wild goat grass, possibly *Aegilops speltoides*. Doubling of chromosome number gave a plant with 28 chromosomes. Modern durum wheat *T. durum* has been cultivated from this plant. Modern bread or hexaploid wheats *T. aestivum* are derived from a further hybridization, between emmer wheat and another goat grass *Aegilops squarrosa*. This added a further 14 chromosomes to bring the complement to 42. The diagrams are ~0.75 life-size.

S. alterniflora to produce the sterile hybrid, *S. x townsendii*, which still persists. In about 1890 a vigorous, fertile polyploid species, called *S. anglica* was derived from the hybrid and has since spread widely around the coast of Britain and Northern France where it has taken over whole areas of salt marsh.

The evolution of the catchfly *Galeopsis tetrahit* which has $2n = 32$ chromosomes has even been experimentally reproduced by artificially crossing two other species, *G. pubescens* and *G. speciosa*, each of which has $2n = 16$ chromosomes. But perhaps the most economically important example of allopolyploidy is the evolution of modern tetraploid and hexaploid wheats (Fig. 1.8), which seems to have resulted from the accidental hybridization of three species of grasses which still grow wild in the Middle East. In fact, many food plants seem to be polyploids of one sort or another; this is probably because the size of plant cells seems to be proportional to the size of their genome. Polyploid plants will therefore have larger cells and so fleshier leaves, larger flowers and bigger seeds and fruits.

1.6.4 Speciation in asexual populations

Many plants, particularly weedy angiosperms, rarely use **sexual reproduction**. Instead they employ **asexual reproduction**, either vegetatively, by producing rhizomes or stolons, or by producing seeds by the process of **apomixis**. In this way they can reproduce faster and colonize new ground more rapidly. This also has the side-effect that each *individual* is reproductively isolated and can change independently from its neighbours. As a consequence, asexual plants such as dandelions and brambles tend to readily form varieties or **microspecies** which are adapted to the local conditions. However, because they rarely indulge in sex, it is very hard to tell whether these are really distinct species.

1.7 MACROEVOLUTION AND EVOLUTIONARY TRENDS

1.7.1 Is macroevolution caused by microevolution?

Despite the relative ease with which Darwinian theory can explain adaptation and speciation, some people still debate whether gradual microevolutionary changes could have produced the great diversity of plant forms. We will probably never be sure if the mechanisms which cause **macroevolutionary** change are the same as those which cause **microevolutionary** change for two reasons: first, because evolution takes place far too slowly for macroevolutionary events to be observed; second, because it is impossible for us to tell now what the major taxonomic groups of the future will be. However, we can glean information from the fossil record about how evolution occurred in the past.

Perhaps the most complete record of the evolution of a major new taxon is the development of mammals from the synapsid reptiles. Throughout the Permian and Triassic periods different groups of these reptiles

gradually changed: developing more complex teeth; larger and better differentiated jaw muscles, a more open skull and a more upright gait. Each of these changes is consistent with the Darwinian concept of gradual adaptive evolution by natural selection; each change would improve the efficiency of food capture and processing, so enabling the animal to raise its activity level and body temperature, and no sudden leaps are seen. This evidence *does* support the view that large-scale evolutionary change occurred by the same mechanism which causes small-scale change.

Unfortunately there is no comparably complete record of the evolution of any major plant group but there is no reason to think that the evolution of plants would occur by a different mechanism from that of animals. Despite this evidence, however, several processes have been suggested which might speed up evolution.

1.7.2 Possible mechanisms for macroevolution

Changes in developmental processes

It has been suggested that large morphological changes can be caused by small alterations in the rate and timing of developmental processes. The speed of development of the body can be speeded up or slowed down, while the onset of sexual maturity can be brought forward or delayed. As a result a new stage can be added on at the end of development, a process called **recapitulation**, or the organism could become sexually mature at an earlier stage of development, a process known as **paedomorphosis**.

Recapitulation seems to be common in animals, whose embryos often seem to retain ancestral characters; for instance, human embryos have gills like fish which evolved much earlier. Paedomorphosis may also be important. Many animal groups may well have been produced by **neotony**, a form of paedomorphosis in which morphological development is delayed. The chordates are often quoted as an example, since they resemble the larvae of their nearest relatives, the tunicates. Paedomorphosis may also be important in plants; herbaceous angiosperms, for instance, may have evolved by accelerating their sexual development, a process called **progenesis**, so that they flower, reproduce and die before the plant has time to become woody.

In a similar way large changes in the shape of organisms could be caused by small changes in the rate of growth of different cell lines. The shape and size of fruit and flowers have been rapidly altered by plant breeders by changing such a mechanism.

Changes in regulatory genes

Another, more controversial, suggestion is that large changes could occur quickly by small alterations in regulatory genes. Such changes in the fruit fly *Drosophila* have been shown to produce flies with four wings, rather than the usual two. Similarly, as we have already seen, single mutations in peas can produce plants with leafy tendrils or with no leaves at all. These **hopeful monsters**, as they are known, could give rise to a whole new group. This is an exciting suggestion but there is little evidence that it is an important cause of change in nature. The main reason for remaining sceptical about this mechanism is that because most organisms are fairly well adapted to their environment, any large changes will almost certainly reduce their fitness. Monsters will rarely survive!

1.7.3 The speed of evolution

There has also been some debate about whether evolution occurred as gradually as Darwin's model suggests. It is becoming clear from the fossil record that species may remain unchanged for long periods of time and that these periods of stasis are interrupted by periods of relatively fast evolutionary change. This **punctuated evolution** model of evolution, however, does not conflict with the Darwinian explanation of change because even during the bursts of rapid evolution the rate of change is really very slow. The rapid evolutionary changes probably coincided with periods when the environment was changing quickly. Such periods seem to have occurred several times in the earth's history, most notably at the mass extinction events at the end of the Permian and Cretaceous periods.

In conclusion, it is possible that various developmental mechanisms may be important in speeding up evolution in certain situations and that evolution varies in its rate. However, there is no real evidence that changes that led to the evolution of major taxa occurred by any different mechanisms from those that cause microevolutionary change. New taxa probably become more important merely because their members tend to survive better, and to speciate more rapidly, than those of other groups.

1.8 LARGE-SCALE TRENDS IN EVOLUTION

Despite the fact that natural selection acts on individual organisms it *is* possible to detect patterns within the evolution of whole groups of organisms (Fig. 1.9). These occur because of the selection pressures

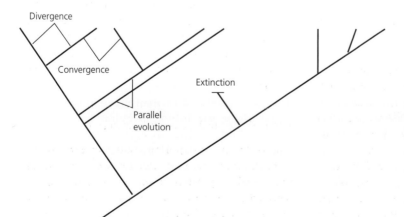

Fig. 1.9 A hypothetical tree of life showing the different patterns of morphological change which are possible in a diversifying group of organisms. After splitting, two taxa can show **divergence**, in which they become progressively more dissimilar, or **parallel evolution**, in which they both alter in the same way. Distantly related organisms can respond to similar selection pressures to become more similar, a process known as **convergence**. Some groups can also become extinct.

under which organisms operate: pressures, which, as we have seen, result from their physical and biological environment.

1.8.1 Divergence

One of the major trends within all groups is **divergence**, in which species with a common ancestry gradually become more dissimilar. This occurs, at least initially, because of competition. Two recently formed sister species will have essentially the same ecological niche, and so competition between the most similar organisms will be most intense. There will therefore be strong selection pressure on both species to become dissimilar. Once they are dissimilar, of course, and inhabit different ecological niches, they will have different biological interactions, and so the selection pressures which drive their evolution will be different. The two species will evolve independently and their form and ecology will further diverge. This is the reason why so many taxonomic groups of plants have members with such widely varying morphology and ecology. The pea family (Fabaceae), for instance, range from tiny clovers to the enormous *Koompassia* trees of tropical rainforests; only when one examines the flowers, whose form tends to be less influenced by a species' ecological niche, can one detect the taxonomic similarity. It is for this reason that the taxonomy of flowering plants is based so strongly on floral characteristics.

1.8.2 Convergence and parallel evolution

A second evolutionary trend which is widely observed is **convergence**, in which distantly related organisms respond to the same selection pressures by developing similar adaptations. There is much convergence between the plants which inhabit the deserts on different continents (see Chapter 10); members of five families, the Cactaceae, Euphorbiaceae, Asclepediaceae, Didieraceae and Apocynaceae have all attained a similar 'barrel cactus' form. Similarly many different species of pioneer trees in rainforests have developed symbiotic relationships with ants which protect them by killing herbivores (see Plate 8, facing p. 110); the species have developed hollow twigs or other structures which shelter the ants and nectaries which attract them. But perhaps even more spectacular, is the convergent evolution of pitcher plants in four different families of angiosperms (Fig. 1.10). They have independently developed liquid-filled pitfall traps which capture and digest insects, providing supplementary nutrients to the plant. The process of **parallel evolution** is essentially similar to convergence, but occurs when closely related species or genera undergo similar patterns of evolution.

1.8.3 Non-convergence

A final evolutionary trend is **non-convergence**, in which distantly related species respond to the same selection pressure by developing *different* adaptations. The 'hedgehog plants' of the Mediterranean, for instance, have responded to drought very differently from cacti, and produce tiny deciduous leaves which are protected within a thorny mat of woody stems. Similarly, many families of angiosperms use quite different mechanisms from the pitcher plants to capture insects; sundews have sticky hairs (see Plate 8, facing p. 112), while the Venus fly trap and its relative *Aldrovandra* (see Chapter 11) have traps produced from closing leaves. In both cases the pattern of evolution has followed a different course because there was a different starting point.

1.9 CAUSES AND EFFECTS OF EXTINCTION

The final factor that has strongly influenced the history of life is **extinction**. Over 99% of all species that have existed have subsequently died out, so the pattern of evolution, like the growth of a topiary bush, is as dependent on what has been cut off as in what remains.

1.9.1 Biological causes of extinction

One of the most important causes of extinction is competition. In nature, the better is the enemy of the good, and so perfectly capable organisms may be replaced by ones that have even a slight advantage in competition. Because competition is fiercest

Fig. 1.10 Convergent evolution of insect-catching pitchers by four species of angiosperms. (a) *Brocchinia reducta* (Bromeliaceae). (b) *Sarracenia flava* (Sarraceniaceae). (c) *Cephalotus follicularis* (Cephalotaceae). (d) *Nepenthes* sp. (Nepenthaceae). The pitchers themselves have been formed either from a number of leaves (a) or from different parts of a single leaf (b, c, d). (All diagrams ×1.)

between organisms with similar ecological niches, species are often outcompeted and replaced by close relatives, perhaps even by sister species which have just evolved. As we have seen, the newly evolved salt-marsh grass *Spartina anglica* is currently outcompeting both its parent species.

In a similar way, the history of evolution has often seen the apparent replacement of one major taxa by another. On land, many plant groups have risen to prominence before being replaced by a new group. In each case it is possible to ascribe this event to some competitive advantage of the new group. For instance, it is often stated that seed ferns replaced club mosses and horsetails because of their more efficient reproduction (see Chapter 7). It is also suggested that seed ferns were in their turn replaced by conifers and angiosperms largely because the latter had a more efficient pattern of secondary growth. Unfortunately, we will never know whether these explanations are correct because we know far too little about the process of competition, even between single species. Frequently, however, members of formerly dominant groups still survive in particularly specialized niches or in isolated regions of the world. The existence of such **living fossils** as the ginkgo (Fig. 7.5, p. 116) always gives us hope that there may remain a 'lost world' where prehistoric plants and animals may still be found.

1.9.2 Environmental change

Another important cause of extinction is undoubtedly environmental change which is often due to climatic shifts. If this happens slowly, plants may have enough time to adapt to the new conditions or may simply alter their distribution. However, fast changes in, or destruction, of island habitats (such as that being caused by humans now both on oceanic islands and rainforests) is likely to result in many extinctions.

1.9.3 Mass extinctions

Perhaps the most pervasive and least understood influences on evolution are the **mass extinctions**, which have occurred at intervals of tens of millions of years. The most famous one occurred 65 million years ago at the end of the Cretaceous period, extinguishing not only the dinosaurs but also many seed plants. Others, however, have been even more devastating; the extinction at the end of the Permian period accounted for 50% of known families of marine animals and 95% of known species.

The causes of mass extinctions are not known with certainty, although meteorites, volcanic activity and supernovas have all been suggested. All three would have global impacts on climate, caused by the scattering of debris in the atmosphere or the extensive release of gases; impacts which could kill organisms all over the world. Clearly when such an unpredictable disaster strikes there will be a strong element of chance in whether a group of organisms survives. It may depend on it producing long-lived seeds or spores or having excellent cold-tolerance. The pruning applied by nature to the tree of life at these points seems to have been essentially random.

1.10 POINTS FOR DISCUSSION

1 Why do you think a large continuous population of plants is unlikely to speciate?
2 What other examples of convergence and non-convergence within the plant kingdom are there?
3 What are the disadvantages of sexual reproduction?
4 How has artificial selection altered (a) crop plants and (b) garden flowers, since they were domesticated?
5 Why were plants better able than dinosaurs to survive destructive events such as meteor strikes?

FURTHER READING

Berg, L.R. (1997) *Introductory Botany: Plants, People and the Environment.* Saunders College Publishing, Fort Worth.
Darwin, C. (1859) *On the Origin of Species by Means of Natural Selection, or The Preservation of Favoured Races in the Struggle for Life.* Murray, London.
Dawkins, R. (1995) *The River Out of Eden.* Wiedenfeld and Nicholson, London.
Price, P.W. (1996) *Biological Evolution.* Saunders College Publishing, Fort Worth.
Ridley, M. (1993) *Evolution.* Blackwell Scientific Publications, Oxford.

CHAPTER 2

The evolutionary history of plant life

2.1 INTRODUCTION

As we saw in the last chapter, today's plants are the end result of hundreds of millions of years of evolution. Therefore if we are to understand the diversity of modern forms we must first reconstruct the pattern of plant evolution and work out how the modern forms are related to each other. In this chapter we will examine how plant taxonomists have used evidence from both living and fossil plants to work out their family tree or **phylogeny**. We will then introduce a pictorial classification of modern plants. This doubles as a short account of the major events in plant evolution and indicates when the major character changes that were developed by the different groups occurred. This information can be used as reference material for the later chapters, in which we will study the adaptations that many different taxa have developed for living in different environments and different habitats.

2.2 THE DIFFICULTY OF RECONSTRUCTING PLANT PHYLOGENY

2.2.1 Using fossils

If we had a complete fossil record of all the species that have ever existed it would be a simple matter to work out the family tree of the plants. All we would need to do would be to trace the way in which they alter and speciate through time, like a botanist tracing the growth and branching pattern of an oak tree. Unfortunately things are not that simple because the fossil record is extremely poor.

2.2.2 Problems with using fossils

Most fossils are formed when dead organisms are preserved by being covered by fine sediment and compressed over a period of millions of years. Four types of fossil may be produced: **impression fossils**, which are the imprint of remains which may have decayed away; **compression fossils**, which are the actual hardened and flattened remains; **cast fossils** which are internal casts of the hollow parts of plants; and **petrifactions** which are very well preserved fossils formed when the plant parts are filled in early on with concentrated mineral solutions.

The major problem with relying on fossils, however, is that fossilization is a very rare event. It only usually occurs around the fringes of stretches of water, and even here only in those rare areas, such as river deltas or coastal swamps, where sediments are being laid down. Because coastlines are always changing, the sediments are rarely laid down continuously, and therefore fossil-containing beds contain many gaps in time or **discontinuities**.

Whatever the process, the rarity of fossilization means that only a very small fraction of the species that have existed have been fossilized. It is also very hard to tell how these species changed over time because of the many discontinuities in the fossil record. Most of the important evolutionary events therefore took place 'off stage' and a great deal of detective work must be carried out to reconstruct them. A large amount of luck is also, of course, involved in actually finding the fossils. A final problem with reconstructing evolution is that it is very difficult to obtain information even from the fossils that we do possess; most clues about their biochemistry and physiology are destroyed by the fossilization process. It also requires great expertise to reconstruct the morphology of complete plants because they break up easily after death into isolated leaves, seeds, twigs, trunks or roots. Palaeobotanists often have to give separate names to each separate organ, and can only rarely find enough evidence to link the different body parts together.

2.3 METHODS USED TO RECONSTRUCT PHYLOGENY

As we have seen, reconstructing the phylogeny of plants and working out the relationship of the living forms is exceptionally difficult. It is like trying to work out what an oak tree looks like when all we have is the tips of the twigs and a few isolated lengths of branch. Despite all these difficulties, however, the relationships of the living plants are fairly well understood, and this is a testament to the ingenuity and hard work of plant phylogenists. They have developed many ingenious techniques to tease out information from living plants and from the few fossils that we have found.

2.3.1 Phenetics

The technique

The most obvious way to work out the relationships of the living plants is to compare their overall similarity; the most closely related plants should be most similar, while more distantly related plants should be more dissimilar. Classifying organisms on their overall similarity in this way is called **phenetics**. To make the concept of overall similarity more objective it is necessary to split each organism into a large number of discrete characters, such as flower colour, leaf length and branching pattern. It is then possible to work out by statistical methods which organisms are the most similar and therefore most closely related, and construct a family tree.

Problems

Unfortunately, this straightforward technique has many disadvantages. First, there are practical prob-

lems. The choice of characters is subjective, and therefore different taxonomists can easily derive quite different family trees from the same organisms. Taxonomies built up from characters taken from different parts of the plant, or from plants at different growth stages can also give different family trees. Even the statistical technique used will influence the answer.

More seriously, there are several biological reasons why this method could produce the wrong family tree. Rather than getting more different through time, some species can actually become more similar, as we saw in Chapter 1, by the process of convergent evolution. A phenetic taxonomist might link the desert-living members of the Euphorbiaceae, which live in Africa, with the Cactaceae of South America because of their overall similarity rather than with their real relatives, the other members of the Euphorbiaceae. The two desert species would be incorrectly linked because they shared **analogies** such as spines —independently derived structures with a similar function—rather than **homologies**, which are derived from the same structure. The spines of euphorbias (Fig. 10.2a) are derived from lateral buds, while those of cacti (Fig. 10.2b, p.165) are modified leaves.

Another problem is that organisms can also evolve at very different rates. For instance, a phenetic taxonomist might decide that oak trees are more closely related to pine trees than to the leafless parasite, dodder, which is a fellow angiosperm. The two trees only resemble each other because they share *primitive* homologies like woody growth. The dodder and oak diverged more recently but since they did so the dodder has changed much more markedly to adapt to its parasitic mode of life.

For these reasons, overall similarity is not a reliable indication of relatedness; both analogies and shared primitive homologies cloud the picture.

2.3.2 Cladistics

The technique

A more precise way of working out phylogeny was developed by the insect taxonomist Willy Hennig. He pointed out that to work out how closely organisms are related the only really useful characters to use are *shared derived* homologies. A good example of a group linked by shared derived characters are the angiosperms, which are linked by common possession of enclosed ovules, bisporangiate flowers and xylem vessels. Just as in phenetics, a branched family tree or **cladogram** showing the relationships of the organisms can then be constructed.

The logic of cladism is indisputable, but the difficulty taxonomists have is that the fossil record is so poor that it is by no means easy to identify *which* characters are homologies and which of these homologies are derived.

Identifying homologies

To determine whether two structures are homologous involves extremely close examination. Homologous organs will have the same fundamental structure, the same relations with other organs, and a similar pattern of development. The flowers of angiosperms clearly satisfy all these conditions, having the same overall pattern, being found at the ends of stems, and forming from the apical meristem in the same way. A final reason for believing flowers to be homologous to one another is that they are extremely complex organs. Such structures are far less likely to have evolved independently several times by convergence than more superficial similarities of organisms in leaf shape or flower colour.

Identifying derived homologies

To determine whether homologous structures are primitive or derived is more difficult, but there are three ways that can be used to help decide. First, you can look at groups of organisms that are more distant relations of those that you are studying, a process called **outgroup comparison**. For instance, when working out the phylogeny of the vascular plants you can tell that the seeds found in seed plants are derived characters because they are not found in the most obvious outgroup, the ferns. However, within the angiosperms, possession of seeds is a primitive character, because the most obvious outgroup, the conifers, also possess them. What is a derived character for a large taxonomic group may therefore be a primitive character when looking at smaller groups.

Of course, you have to possess some idea of the pattern of evolution before you start this process because you have to be able to identify an outgroup. There is therefore a certain degree of circularity in the logic of outgroup comparison.

A second method is to examine the pattern of development of the structure. Primitive characters tend to appear earlier in embryology than do derived ones. For instance mammalian embryos have gills like fish early in their development. In just the same way you can use fossil evidence; primitive characters should appear in the fossil record earlier than derived ones. Fish with gills appear much earlier in the fossil record than mammals. Unfortunately, neither of these methods is totally reliable. For example the fossil record is so poor that organisms with the advanced character state may be found earlier in the record than organisms with the primitive state. The relationships of many families of angiosperms is obscure for just this reason; the fossil record of their early evolution is too poor to base any conclusions on it. A final problem is that it is difficult to date fossil beds.

Resolving the difficulties

As a result, in many cases it is impossible to tell whether characters are really shared derived homologies, even after they have been extensively studied. Therefore, taxonomists often have to use statistical techniques to help identify evolutionary relationships. In such cases taxonomists choose the evolutionary tree that has involved the *least* amount of convergence and so the *least* total amount of evolution. This is called the principle of **parsimony**. A simple example of the use of cladistics in working out relationships between plants is given in Box 2.1.

2.4 TYPES OF EVIDENCE

2.4.1 Morphology

Morphological characters such as those we examined above have been, and continue to be, extensively used to work out the evolutionary relationships of plants. They have the advantage that they are cheap and easy to study. Aspects of plant morphology

BOX 2.1 USING CLADISTICS: A CASE STUDY

This example is an investigation of the relationships between three plants: the rowan *Sorbus aucuparia*, the locust tree *Robinia pseudacacia* and white clover *Trifolium repens.* Table 2.1 shows five characters that are possessed by one or more of these species and by another less closely related species the magnolia. This acts as the outgroup. Not all of these five characters will be useful in helping us to determine their relationships. All four species have xylem vessels, and therefore this character cannot tell us anything about their interrelationships. Similarly, although only two test species, the rowan and the laburnum, are woody, this character is also useless for this purpose. This is because the outgroup, the magnolia, is also woody, telling us that this is a primitive character. The clover must have evolved herbaceousness, rather than the two tree species evolving woodiness. There are therefore just three characters remaining which are both derived characters and which are also shared by two test species. The locust tree and clover share asymmetric pea flowers and root nodules, while the rowan and locust tree share pinnate leaves (Fig. 2.1). These facts suggest that two

Table 2.1 Phylogeny of rowan, locust tree and clover. Five characters are shown. The only characters that are useful in constructing the tree are **shared derived characters**. These are identified as characters that have changed from the ancestral state, which is seen in the outgroup plant, the magnolia. The locust tree and clover share pea flowers and root nodules, whereas the locust tree and rowan share pinnate leaves.

	Species			
Character	*Magnolia*	*Rowan*	*Locust tree*	*Clover*
Xylem vessels	+	+	+	+
Woody stem	+	+	+	–
Pea flowers	–	–	+	+
Root nodules	–	–	+	+
Pinnate leaves	–	+	+	–

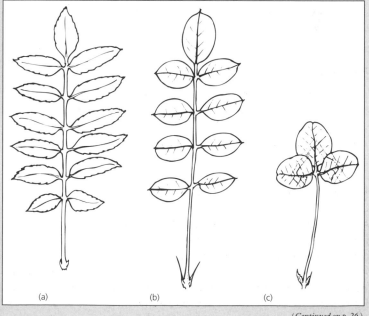

Fig. 2.1 Convergent evolution in leaf form. The rowan has pinnate leaves (a) similar to those of the locust tree (b). However, analysis of other characteristics shows that the locust tree is more closely related to clover which has trifoliate leaves (c). It is generally accepted that vegetative characters of plants are more plastic than reproductive ones, so the fact that the locust tree and pea both have similar pea flowers is used as a more reliable indication of relationship. (All diagrams ×0.4.)

(*Continued on p. 26.*)

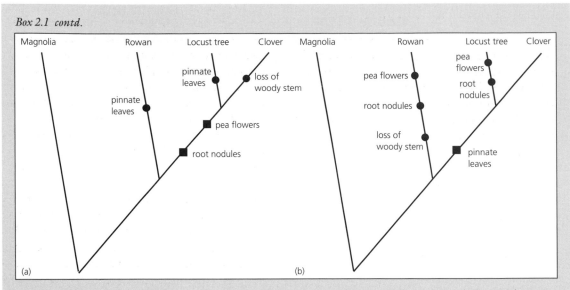

Box 2.1 contd.

Fig. 2.2 Two possible evolutionary trees for the four species described in Table 2.1. The preferred choice is (a) because the locust tree and clover share **two** derived characters, whereas the locust tree and rowan (b) share only **one**. Tree (a) therefore involves less convergence and fewer evolutionary steps, five rather than six in (b). Squares denote shared derived characters, circles independently derived characters.

alternative evolutionary trees or cladograms are possible; either the locust tree and clover could have diverged most recently (Fig. 2.2a) and therefore be most closely related, or the locust tree and rowan (Fig. 2.2b). Neither cladogram explains all the evolution, so at least one of these changes must be the result of convergent evolution.

In the case of the plants given here, the first tree involves only a single instance of convergence: the evolution of pinnate leaves, while the second involves two: the evolution of pea flowers and root nodules. The first tree is therefore chosen by taxonomists to be the most likely one as it involves a total of only five rather than six evolutionary steps. This means that the rowan and locust tree would have acquired pinnate leaves by convergence. This is perfectly possible; a survey of the angiosperms shows us that leaf shape is indeed very plastic and is far more likely to occur by convergence than the evolution of the complex pea flower or root nodules.

Similar techniques using the principle of parsimony can be used to investigate the evolutionary relationships of much larger numbers of organisms. However, the complexity of the calculations soon becomes unmanageable and computers have to be used to determine which of the huge number of possible trees is most likely.

can also be readily related to function so it is fairly easy to recognize cases of convergent evolution. Morphological characters have the further advantage that they are well preserved in fossils and so it is relatively easy to determine which state is the derived one.

One major problem with using morphological characters is that they tend to show a great deal of convergence. As we shall see throughout this book, this is particularly true of the vegetative systems of plants; natural selection seems to have repeatedly tested out good ideas, such as the development of multicellularity, differentiation, secondary thickening and xylem vessels. Experience has taught plant taxonomists that the reproductive system tends to be far less plastic than the vegetative part of plants and so they mainly rely on reproductive characters to produce their classifications. This is why angiosperms tend to be classified on the basis of their floral structure, members of the same family usually having similar flowers, even if their vegetative parts look totally different.

2.4.2 Biochemistry and cytogenetics

There are real limitations to the traditional morpho-logical methods, however. They require huge skill and expertise on the part of the taxonomists who apply them, and it is difficult for them not to be sub-jective. In addition many of the simplest organisms such as the unicellular algae and some parasites have very few obvious morphological characters that can be used. To make up for these defects, information has increasingly been used from different sources. One such source is the structure and biochemistry of the chloroplast and the cell wall. Another is the num-ber, size and structure of the chromosomes.

2.4.3 Molecular sequence data

Increasingly, information is being obtained from molecular evidence: the sequences of the proteins produced by cells and the sequences of the DNA and RNA which code for them. These techniques have several advantages over traditional methods. First, it is easy to recognize discrete characters; they are simply the different amino acids or bases in the sequence. Second, molecular sequences provide sub-stantial information because they are extremely long. Third, there is no danger of being subjective.

A further advantage is that different molecules can be chosen to investigate evolution depending on the time scale the investigator wishes to examine. Mitochondrial DNA mutates faster than other sorts and so can be used to investigate the evolutionary relationships of closely related species and genera, which have diverged in the last few million years. In contrast, ribosomal RNA mutates very slowly, over periods of hundreds of millions of years, and so has recently been used to elucidate the relationships between the major taxa of organisms. Nuclear DNA and the DNA of certain conservative chloroplast genes alter at an intermediate rate and can be used to investigate taxa of intermediate size. The *rbc*L chloroplast gene (a large subunit of the ribu-lose-1,5-biphosphate carboxylase gene) has recently been used to reassess the relationships of the seed plants.

However, despite these many advantages, molecu-lar sequences also have certain disadvantages. First, it is so difficult to relate the molecular sequence of a protein to its function that it is virtually impossible to identify convergence. Molecular taxonomists are therefore more heavily reliant on their computer software to work out phylogeny. Second, these tech-niques can only be used to investigate living plants because DNA is almost always destroyed by fossiliza-tion. The techniques of molecular biology are also still rather expensive. Third, there is the problem of **multiple base exchange**: if, for instance, a base A changes to C then back to A no evidence of evolution will be found at that site. Fourth, if groups that have split a long time ago are compared, and no intermed-iate forms are known, they often look more closely related than they are. This problem is called **long branch attraction**. Finally, just as classifications based on morphological characters from different parts of the body can be different, so can those based on the study of different molecules or even parts of molecules. To overcome this difficulty, a final method that investigates the overall similarity of the DNA from different species has been developed: a method called **DNA hybridization.**

2.4.4 DNA hybridization

In this technique, DNA from two different species is put together and heated. The double strands melt, producing single strands each of which may reform when cooled again, or may join up with a homol-ogous strand from the other species. The hybrid strands will not be so stable as the single species strands because their base pairs will not all join up; consequently their melting points will be lower. The phylogeny can be worked out by comparing the melting points of hybrid and single species strands; the more similar the species the more stable the hybrid DNA and therefore the more similar the melting points. This straightforward technique has proved extremely useful. However, because it exam-ines overall similarity of the DNA, just as a phene-ticist examines overall morphological similarity, it does make one major assumption: that DNA changes at approximately the same rate in all species. Whether it does so is far from certain; herbaceous plants, for example, seem to have a faster clock than longer-lived woody plants, and therefore results obtained by using this so-called **molecular clock** must be used with caution.

2.4.5 Pooling information from different sources

As we have seen, morphological and molecular techniques both have advantages and disadvantages, so nowadays plant taxonomists are sensibly combining results from both techniques to work out the phylogeny of plants. Our knowledge of the pattern of evolution is continually improving. However, despite all their best efforts taxonomists will never be able to be completely sure of the phylogeny of living plants. Certain groups may have evolved over too short a period of time and too long ago to leave us with enough information; there may be few, if any shared derived characters and their molecular sequences will have changed too much since they evolved. The place of fossil forms is even harder to determine. Nevertheless, the success of taxonomists is impressive and continues to improve.

2.5 SYSTEMS OF CLASSIFICATION

Once the phylogeny of plants is known and the pattern of evolution has been elucidated it is then possible to classify them. However, even if the phylogeny is known, many different classifications are still possible.

2.5.1 Cladism

An influential school of taxonomists called the cladists believe that the most logical way of classifying organisms is for the classification to exactly follow the phylogenetic tree. In the cladists' opinion, only **natural groups** which are linked by shared derived characters should be classified together. The problem with using this approach strictly is that the system produces large numbers of groups, each of which has to be named. Moreover, the groups are all too likely to change as new classifications are produced.

Another great difficulty is that most existing classifications have not been carried out in this way, largely because they were worked out before the logic of cladistics, or even ideas about evolution, were accepted. The groups most widely known to the general public, and to most botanists, are those such as the algae, bryophytes and gymnosperms, which are grouped essentially on their overall similarity. Many of these groups are linked by shared primitive characters and are essentially 'rag, tag and bobtail' groups of organisms with no close relationship at all. The algae, for instance, are a disparate group which includes all the photosynthetic eukaryotes, except for the land plants. Many algae are more closely related to heterotrophic organisms than other algae. For example, the brown algae and diatoms are more closely related to the fungi-like oomycetes than to red or green algae.

The final problem is that cladistic classifications cannot easily be shoehorned into strict hierarchical classification, in which species are clustered into a genus; genera into a family; families into an order; orders into a class; classes into a phylum; and phyla into a kingdom. Such groups reflect our own needs and patterns of thinking rather than the process of evolution.

2.5.2 Evolutionary taxonomy

For these reasons another group of botanists, the evolutionary taxonomists use a rather more pragmatic and subjective approach. They make use of the findings of cladists, but also use information about overall similarity to produce a sort of compromise classification. Their classifications are logically flawed but usually easier to follow than those of cladists.

2.6 A SIMPLIFIED CLASSIFICATION OF PHOTOSYNTHETIC ORGANISMS

To overcome the problems encountered by conventional classifications of plants we have decided that it is perhaps most useful to present just the phylogenetic tree of plants, with some indications about how groups have been linked in conventional classifications. Figs 2.3–2.5 therefore are a series of

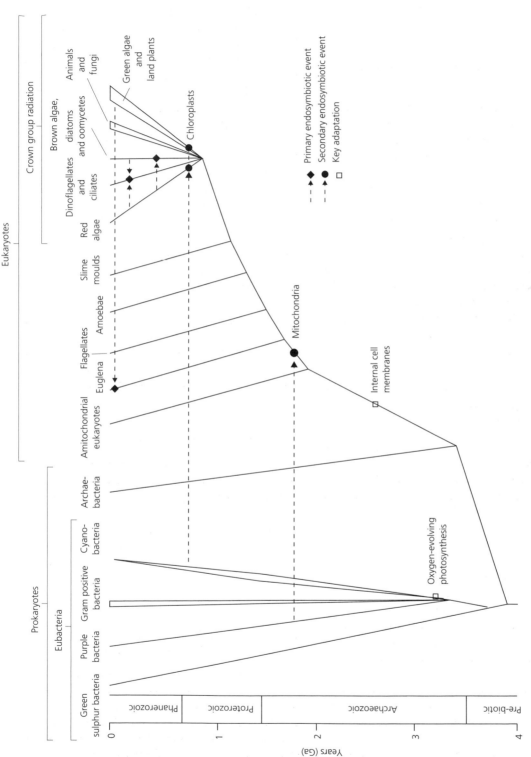

Fig. 2.3 A universal family tree of life, showing when the main groups of organisms evolved and their relations with each other. Life first developed around 4 billion years ago, and two major groups of prokaryotes evolved: the Eubacteria and the Achaebacteria. Eukaryotes evolved around 2 billion years ago from relatives of the Archaebacteria, following several developments: the production of internal cell membranes; and the capture of purple bacteria, to form mitochondria, and of cyanobacteria, to form chloroplasts, a process called endosymbiosis. Several groups of photosynthetic organisms have since acquired chloroplasts by secondary endosymbiosis of photosynthetic cells. The major eukaryote groups evolved and split apart from each other around 1 billion years ago, giving rise to many groups of algae, plants, fungi and animals. This is known as the **crown group radiation**. Endosymbiotic events are shown as horizontal arrows linking branches on the tree.

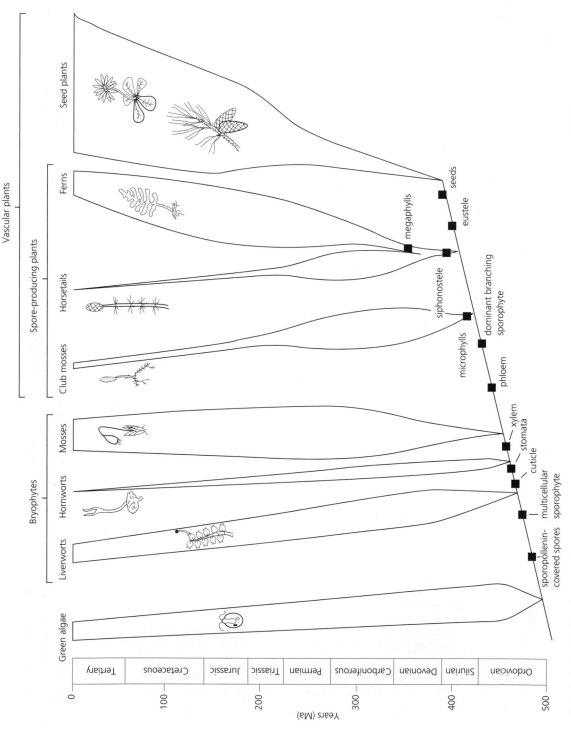

Fig. 2.4 A simplified family tree of the land plants, showing when the main groups evolved, their relationships with each other, and their relative success (shown as the width of the lines). Key adaptations (■) made by each group are shown along the tree. Conventional classifications split the land plants into three groups: bryophytes, in which the gametophyte is dominant; spore-producing vascular plants, in which the sporophyte is dominant; and seed plants, in which the female gametophyte is held within the seed.

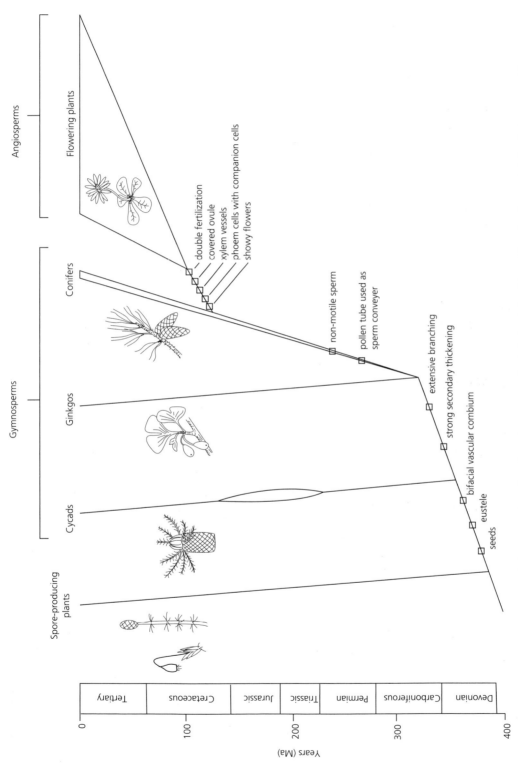

Fig. 2.5 Simplified family tree of the seed plants, showing when the main groups evolved, their relationships with each other, and their relative success (shown as the width of the lines). Key adaptations (□) made by each group are shown along the tree. Conventional classifications split the seed plants into two groups: the gymnosperms, which have a partly exposed ovule, and the angiosperms in which the seed is completely enclosed.

Within the figure the following labels appear: Angiosperms; Gymnosperms; Spore-producing plants; Flowering plants; Conifers; Ginkgos; Cycads; double fertilization; covered ovule; xylem vessels; phloem cells with companion cells; showy flowers; non-motile sperm; pollen tube used as sperm conveyer; extensive branching; strong secondary thickening; bifacial vascular combium; eustele; seeds. Time scale (Years (Ma)): 0, 100, 200, 300, 400; Tertiary, Cretaceous, Jurassic, Triassic, Permian, Carboniferous, Devonian.

cladograms, which show the phylogenetic tree of photosynthetic organisms. This has been reconstructed using the best evidence currently available, including the most recent fossil finds and molecular data from ribosomal RNA, and mitochondrial and chloroplast DNA sequences. This arrangement allows one at a glance to see the evolutionary relationships of the plants far more easily than using a conventional classification. To help readers who are used to the traditional approach, the way in which the different groups have in the past been fitted into classes is mapped out above the tree.

These diagrams can also be used to help tell the story of the evolution of plant life as it includes other information. The widths of the lines gives some indication of the relative success of the different groups. A geological time scale on the left of each figure also gives information about when the different groups evolved. Finally, the more important evolutionary changes that occurred, many of which are **key adaptations** opening up totally new ways of life, are marked along the tree. Together with the work of palaeoclimatologists in reconstructing the environments of the past this enables us to put together a coherent story of plant life which explains to a certain degree *why* evolution followed the pattern it did. As we shall see throughout the book, the relative success of the different groups can be related in large part to the key adaptations that they developed.

A list of plant groups is also given at the end of this book, together with information about their morphology, biochemistry, ecology and diversity.

2.7 POINTS FOR DISCUSSION

1 Why do you think it is important to know the phylogeny of plants?

2 Why do you think a classification based on phenetics will be less useful than one based on cladistic principles?

3 In the light of the acquisition by plants of mitochondria and chloroplasts by endosymbiosis, and of plant hybridization and allopolyploidy, do you think a branching classification is valid for plants?

4 Will we ever be sure of the evolutionary relationships of plants? If not why not?

5 Do you think that the traditional classification of organisms into hierarchical groups such as genera, families, orders and classes is biologically meaningful or useful?

FURTHER READING

Bhattacharya, D. & Medlin, L. (1998) Algal phylogeny and the origin of land plants. *Plant Physiology* **116**, 9–15.

Futuyma, D.I. (1998) *Evolutionary Biology*, 3rd edn. Sinauer Associates, Sunderland, MA. [Read Chapter 5.]

Kenrick, P. & Crane, P.R. (1997) The origin and early evolution of plants on land. *Nature* **389**, 33–39.

Niklas, K.J. (1997) *The Evolutionary Biology of Plants*. University of Chicago Press, Chicago.

Price, P.W. (1996) *Biological Evolution*. Saunders College Publishing, Fort Worth. [Read Chapter 13.]

Stewart, W.N. & Rothwell, G.W. (1993) *Palaeobotany and the Evolution of Plants*, 2nd edn. Cambridge University Press, Cambridge.

Part 2

Life afloat

3.1 INTRODUCTION

Because this book is an account of the diversity and evolution of the plants, we must first define what a plant is. Plants are **autotrophic** organisms which make their own organic compounds using energy from the sun, by the process of **photosynthesis**. This ability to make their own food differentiates them from **heterotrophic** organisms such as animals, fungi and many bacteria, which rely on organic compounds in their environment.

This chapter will consider what we can deduce about the earliest plants, examine the long evolutionary history of life afloat, and describe the single-celled plants that continue to jostle for survival in our oceans, rivers and ponds.

3.2 THE FIRST PLANTS

We know almost nothing for certain about the very first plants, but we can make educated guesses about them based on the evidence of fossils, molecular clocks and present-day organisms. Plants probably originated around 3.5 billion years ago in an aquatic environment (the sea was nowhere near as saline at this time), when simple prokaryotic unicells acquired the ability to make organic material from inorganic building blocks with the help of sunlight.

3.2.1 Bacteria using unconventional photosynthesis

Some **archaebacteria** developed a means of trapping light energy using **bacteriorhodopsin**, a pigment not unlike our own visual pigment. The energy was then used to pump protons out of the cell. This process yields little energy, however. The few **halophytic** bacteria that use this process today are restricted to salt pans where few competing organisms can survive.

Much more efficient methods of trapping light energy were developed by the **eubacteria**. In these organisms light energy is first absorbed by the green pigment **chlorophyll**, which acts together with one or more accessory pigments. This energy is then used to reduce carbon dioxide to form sugars. Several different forms of bacteria share this basic mechanism of **photosynthesis** but use different oxidizable substances to reduce the carbon dioxide. The photosynthetic purple non-sulphur bacteria use a range of organic compounds such as alcohols and fatty acids as electron donors. Two other groups, the purple and green sulphur bacteria, instead use hydrogen sulphide in the reaction:

$$6CO_2 + 12H_2S \rightarrow C_6H_{12}O_6 + 6H_2O + 12S$$

The distribution of these organisms is, however, once again limited by the rarity of the sulphur to hydrothermal vents and sulphur-rich lakes.

3.2.2 Evolution of conventional photosynthesis

The group of organisms that proved most successful, the **cyanobacteria**, developed the ability to use the far more common substance, water, as the electron donor in the familiar conventional photosynthesis reaction:

$$6CO2 + 12H_2O \rightarrow C_6H_{12}O_6 + 6H_2O + 6O_2$$

Because they release oxygen as a byproduct of this reaction, the cyanobacteria in time started to alter the earth's atmosphere. At first, much of the oxygen was used up oxidizing the ferrous iron which was common in the earth's crust to form red ferric deposits, which later formed rocks. However, this process seems to have been complete by around 1.8 billion years ago, and an oxidizing atmosphere started to form. This change had two very positive effects on life. First, the oxygen gave rise to the ozone layer in the upper atmosphere which shields life from much damaging ultraviolet radiation. Second, the oxygen could be used to break down sugars, in the process known as **respiration**, into carbon dioxide and water:

$$C_6H_{12}O_6 + 6O_2 \rightarrow 6CO_2 + 6H_2O$$

Respiration is the reverse process to photosynthesis, and therefore life could be based on an efficient re-cycling scheme in the reactions:

$$6CO_2 + 6H_2O \leftrightarrow C_6H_{12}O_6 + 6O_2$$

The evolution of respiration would have led to further increases in the productivity of the cyano-bacteria. These steps, as with many we will examine in this book, prompted an explosion of different types of photosynthetic organism, some of which still thrive in aquatic habitats today. As we shall see, how-ever, the cyanobacteria, and perhaps another recently discovered group of photosynthetic bacteria, the **Prochlorophyta**, not only remain successful organ-isms, but also form an important component of all other plants.

3.3 THE CYANOBACTERIA (CYANOPHYTA)

3.3.1 Structure

The simplest cyanobacteria alive today may resemble the first oxygen evolvers, which were undoubtedly prokaryotes. Like all other plants, as we mentioned above, they owe their ability of photosynthesis to the pigment systems that can capture the energy from light and use it to make organic compounds from inorganic molecules. Present-day cyanobacteria possess **chlorophyll *a*** (as in all green plants) but the molecules that capture light energy and bring it into the reactions of photosynthesis are **phycobilins** (as in very few photosynthetic organisms). The latter are molecules rather like bile pigments, which can have a profound effect on the colour of the cells. The group is sometimes known as the blue–green bacteria (or blue–green algae, or even just 'the blue–greens') but they are by no means all blue–green in colour. One of the phycobilins, phycoerythrin, is red, and a pre-ponderance of this produces red-coloured cells.

These simple bacteria, like all plants, carry out photosynthesis in specialized parts of their cells. For both cyanobacteria and land plants these regions are systems of membranes called **thylakoids** (Fig. 3.1). Cyanobacterial cells contain no other internal mem-branes. The remaining contents of these simple cells include a loop of DNA in the central region, RNA and reserves of various kinds, but no membrane-bound compartments. The particles embedded in the thylakoids seen in Fig. 3.1 are phycobilins, but these membranes are otherwise very reminiscent of those seen in the chloroplasts of land plants.

3.3.2 Growth form

Unicellular cyanobacteria

The majority of cyanobacteria are unicellular. They divide by a simple process very like that seen in ani-mal cells, in which the two ends of cells are 'pinched off' to form two separate daughters. Apart from their rigid outer cell wall, the simplest of these cyano-bacteria have few other structural features that con-tribute to their success. Partly this is due to their small size. Most are so small, being less than 2 μm in diameter, that they have a high surface to volume ratio and therefore nutrient uptake is very efficient. This helps them tolerate very low nutrient levels, and they can form the dominant life form of the cool, dim inhospitable waters of subtropical oceans 50–100 m beneath the surface. They do not sink, partly because they are only slightly denser than water, and partly because they are so tiny; the speed at which spherical objects sink is proportional to

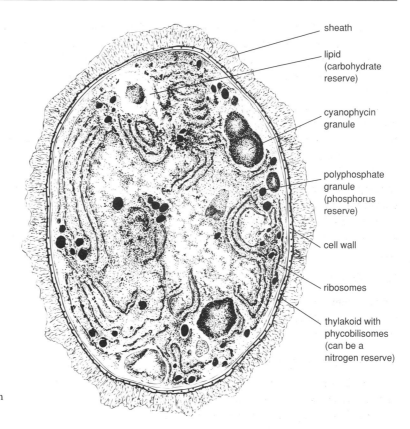

sheath

lipid
(carbohydrate
reserve)

cyanophycin
granule

polyphosphate
granule
(phosphorus
reserve)

cell wall

ribosomes

thylakoid with
phycobilisomes
(can be a
nitrogen reserve)

Fig. 3.1 Drawing (×40 000) made
from a transmission electron
micrograph of a section through a
cyanobacterial cell. *Anabaena* is a
filamentous species, and the sheath
around such cells is especially
prominent, but the contents include
structures and reserves that are found
in most cyanobacteria; the most
conspicuous inclusions are reserves.
The only membranous structures are
the thylakoids; there is no nucleus and
the DNA is found free in the cytoplasm
towards the centre of the cell.

their effective density and to the square of their
radius.

Filamentous cyanobacteria

There are also a large number of filamentous
cyanobacteria, whose cells remain attached after division. We can only guess at how **multicellularity**
evolved. Perhaps it was due to the failure of complete
separation of daughter cells, but it proved to be a
critical step in the evolution of complex plants, as we
shall see in the next chapter. Many filamentous
cyanobacteria are free-floating **planktonic** organisms (Fig. 3.3). Others form **microbial mats** on the
surface of sandy or muddy sediments. Since such
mats trap sediment, new layers are gradually laid
down and many-layered rocks or **stromatolites** (Fig.
3.2) are formed. Stromatolites around 3.1 billion
years old are among the first authenticated fossils
on earth and they were once very common. They

declined, however, once grazing protozoans and animals evolved at the end of the Precambrian period.
Present-day stromatolites are restricted to very saline
areas, such as Shark Bay in Western Australia, where
protozoans and other small herbivorous animals cannot survive.

Although cyanobacteria lack flagella to propel
them, many filamentous forms are able to move.
This ability to transport entire multicellular plant
bodies around is unique to cyanobacteria and puts
species with this capability at a distinct advantage in
sandy sediments; the filaments can raise themselves
to the surface if they are buried and they can even ori-
entate themselves to the sun. Just how the cells or
filaments move is not well understood, but tiny
strands (microfilaments) associated with their cell
walls and the action of compounds known to
influence muscular contraction indicate that the
mechanism may be similar to the actin–myosin system of animals.

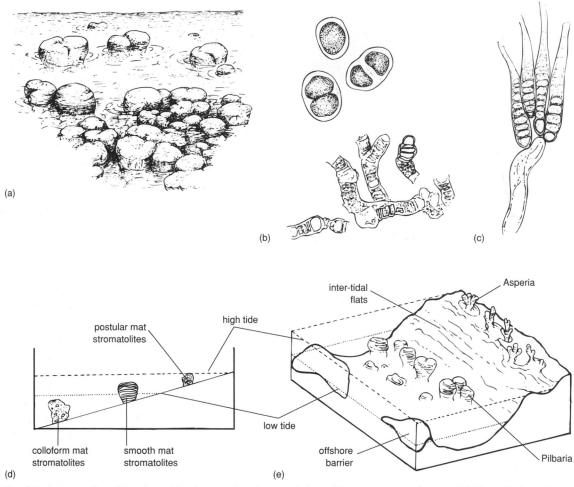

Fig. 3.2 Stromatolites. Mats formed by the growth and accumulation of filamentous cyanobacteria. (a) View of a shoreline (×0.01) at Hamelin Pool, Western Australia, showing stromatolites exposed at low tide. (b, c) Types of cyanobacteria (×5000) that form stromatolites here. (d) Variation of stromatolite form with depth. (e) Reconstruction of environment in which stromatolites grew in the Pilbara region 2 billion years ago. The morphology of these stromatolites is almost identical to modern forms.

3.3.3 Metabolism

Carbon metabolism

The morphological simplicity of cyanobacteria hides enormous metabolic plasticity. Some living species are capable of using hydrogen sulphide, rather than water, as an electron donor in photosynthesis, and can switch from one type of photosynthesis to the other as environmental conditions dictate. A tolerance of extremes of temperature coupled with metabolic plasticity, such as an ability to use organic forms of carbon in the dark, also helps cyanobacteria inhabit some of the most hostile and remote parts of our planet (see Chapter 10). A tolerance of extreme shade also gives some species a big advantage over more conventional photosynthesizers, and they are important members of the **picoplankton**, the small floating organisms of the world's oceans.

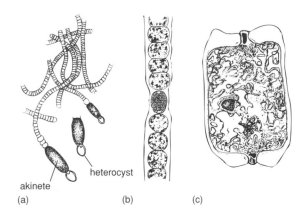

heterocyst

akinete
(a)   (b)   (c)

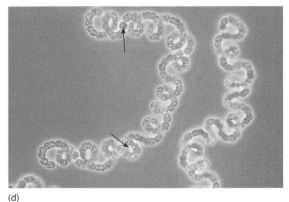

(d)

Fig. 3.3 Division of labour in cyanobacteria. (a)
Cylindrospermum (×500), a cyanobacterium capable of
fixing atmospheric nitrogen in their terminal **heterocysts**.
The other unusual-looking large cells are **akinetes**:
reproductive bodies explained in section 3.4. (b) *Anabaena
flos-aquae* (×3000), which has centrally-positioned
heterocysts. These are separated off from the surrounding
cells by a very thick wall and are only connected to the
neighbouring cells via tiny channels at either end. These
features are easier to see in the drawing taken from an
electron micrograph (c). (d) Light micrograph of *Anabaena
spiroides* (×1000) with the heterocysts visible within the
strands (see arrows).

Nitrogen metabolism

Similar plasticity is shown in their nitrogen
metabolism. The thylakoids of some cyanobacteria
start to resemble those of land plant chloroplasts
more closely if the cells are starved of nitrogen. The
photosynthetic biliprotein pigments are rich in nitrogen and they can be broken down to supply the

nitrogen needs of the cell. This switches the cells into
a less efficient mode of photosynthesis, but is yet
another illustration of the ability of these organisms
to adapt to changes in the environment. The experts
in the group, however, do not need to resort to this
strategy as they can fix nitrogen from the air. This is
something no other group of photosynthetic organisms can manage, and puts the cyanobacteria capable
of nitrogen fixation at a powerful advantage over
the others in nitrogen-poor conditions. There is
one snag with such fixation, however, which is that
the operation of the critical enzyme, nitrogenase, is
prevented by oxygenated conditions. The fixation of
carbon in photosynthesis generates oxygen, so the
two processes cannot occur at the same time in the
same place.

Different cyanobacteria have come up with various
ways of overcoming this problem. Some manage
to photosynthesize at certain times then switch to
nitrogen fixation at others. The most complex multicellular cyanobacteria, in contrast, separate the two
spatially. They have specialized cells called **heterocysts** (Fig. 3.3), which, lacking a photosystem essential for photosynthesis, evolve no oxygen. Such cells
depend on their neighbours for organic forms of carbon, but give them nitrogen in a form they can use in
return. This specialization, in which particular cells
are used for particular purposes, is a form of **division
of labour** and the 'blue–greens' are the only bacteria
that show this characteristic. Rather than being just a
haphazard mass of single cells, they are truly ordered
interconnected systems of cells.

3.3.4 Reproduction

Simple cell division (Fig. 3.2b) is the commonest
mode of reproduction in cyanobacteria, but some
species have various alternatives. Two of these methods of reproduction confer advantages in changing
conditions, such as when nutrients or water become
limiting. Both rely on the formation of structures
called **spores**, which have specially thickened walls,
capable of resisting desiccation and allowing cells to
sit out unfavourable periods in a state of suspended
animation. This is the trick behind the almost magi-
cal appearance of cyanobacteria in pools of water that
collect in the desert just a few hours after a shower of

rain. They germinate and grow during the brief periods when moisture is available, forming more thick-walled resistant bodies as it vanishes, ready for years or even decades of inactivity. Some species form several spores inside each cell, but some filamentous species generate special enlarged cells called **akinetes**, which are packed with DNA and reserves and enveloped by a specially thickened wall. The akinetes are all that survive when the parent filament dries up or dies (where they can lurk, for example, at the dry soil surface, until conditions moisten).

Another mode of reproduction seen in filamentous cyanobacteria involves the separation of some parts of the filament from others. At its simplest this is merely fragmentation; at its most sophisticated it involves the formation of **hormogonia**. These are short chains of cells (Fig. 3.4a–c) which in motile species are the most mobile, and so effect distribution of the organism to new sites.

Prokaryotes do not reproduce sexually, but cyanobacteria, like their non-photosynthetic relatives, are not wholly reliant on mutation (see Chapter 1) to generate genetic variation. Viruses can penetrate and enter bacterial cells, and 'infection' may include the transfer of DNA from one bacterium to another. The viruses that specialize in cyanobacteria are called cyanophages, and are thought to be one method of transfer of the small plasmids of DNA which can introduce new genes. There is also some evidence that transformation and conjugation occur

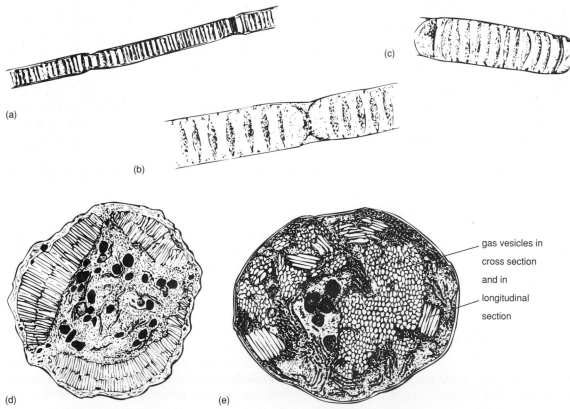

Fig. 3.4 Reproduction in cyanobacteria. (a–c) Reproduction by the formation of **hormogonia**. (a) A filamentous cyanobacterium (×2000) before hormogonia formation. (b) Separation between two cells creates a weak point at either end of the hormogonium, which eventually detaches from the parent filament and then acts as a reproductive body (c). (d, e) Drawings from transmission electron micrographs (both ×25 000) of *Nostoc* (d) and *Trichodesmium* (e), two cyanobacteria that have 'gas vacuoles' (circled). *Nostoc* has its gas vesicles in bands around the periphery of the cells, while those of *Trichodesmium* are less orderly. Each 'vacuole' comprises many cylindrical vesicles, seen in both cross and longitudinal section in *Trichodesmium*.

BOX 3.1 CYANOBACTERIAL BLOOMS

Cyanobacterial blooms occur when conditions are ideal for their growth, when their ability to reproduce rapidly causes a huge build-up in their numbers. The need to control such blooms comes from their ability to produce compounds harmful to humans and livestock. Blooms form naturally in many conditions. Some occur as a result of upwellings of nutrient-rich waters in the sea at certain times of year, for example. It is the blooming of certain red cyanobacteria which gives the Red Sea its name. However, blooms also form when environments are perturbed by the activities of humans. An example is the thick mat of cyanobacterial cells that can form at the surface of ponds and lakes subjected to artificially nutrient-rich run-off from farms. Such mats can cause dermatitis and a range of more unpleasant effects in people and animals which come into contact with or drink the water, but can also end up choking the life out of the water. The cyanobacteria, together with other bloom-forming eukaryotes, reduce the light reaching other photosynthetic organisms within the water. Oxygen levels fall, bacterial decay sets in, and the water can end up so anaerobic it becomes incapable of supporting plant or animal life. Cyanophages or larger cyanobacterial pathogens are currently being sought in an attempt to provide a biological agent capable of controlling such blooms without the need to resort to chemicals.

Surface blooms of cyanobacteria testify to their ability to reach the best-illuminated conditions, which in aquatic situations does not reflect muscular-type movement. In the most sophisticated cyanobacteria the credit lies with structures called **gas vacuoles** (Fig. 3.4d, e). These consist of assemblages of tiny cylindrical structures called **gas vesicles**. These structures act like floats to cause the algae to rise to the surface, but despite their name, however, they are not filled with gas under pressure, like balloons. Rather, they resemble pressurized submarines, merely excluding water. This means that if the pressure of the cytoplasm around them reaches a particular critical level, they collapse, and the buoyancy they were conferring on the cell is lost. The most powerful influence on the osmotic pressure exerted by the cytoplasm is the organic carbon compounds made during photosynthesis. Thus, in very high levels of light, photosynthesis is rapid, and sugars accumulate. Gas vesicles collapse, the cells sink down, photosynthetic rate is reduced, new gas vesicles form and survive, the cells rise, and so on. This mechanism allows cyanobacteria with gas vesicles to optimize their position in the water column, to the detriment of their competitors.

in cyanobacteria, but the processes are far better documented in bacteria. The genetic variation these processes confer causes us problems when we attempt to prevent cyanobacterial blooms (see Box 3.1).

3.4 THE EVOLUTION OF EUKARYOTES

All the eukaryotic plants are quite different from prokaryotes such as cyanobacteria. Their cells are generally much larger, being over 10 µm in diameter. They are also more complex structures, having internal division of labour, and being split up into several compartments by internal membranes. They contain several organelles: the nucleus, mitochondria, flagella and, in photosynthetic cells, chloroplasts (see Figs 3.5–3.7). DNA is found not only in the nucleus, but also associated with some of the other organelles.

3.4.1 Heterotrophic eukaryotes

Heterotrophic eukaryotes first appeared around 1.5–1.3 billion years ago. Studies involving electron microscopy and analysis of the gene sequences of the organelles strongly suggest that at least some of these changes resulted from **endosymbiosis** of enlarging cells with prokaryotes. The first stage in the evolution of eukaryotes seems to have been capture by the cell of respiring purple bacteria. These subsequently remained within the cell and evolved into energy-producing **mitochondria**.

The second stage may have been the capture of

long thin bacteria, like the spirochaetes found in the guts of ruminants. These would have evolved into the motile **flagella** and furnished the microtubular apparatus which is used in mitosis and meiosis. The third stage, the development of a nucleus, was probably not achieved by endosymbiosis, however. The nuclear membrane which wraps up the main body of DNA was probably formed by infolding of other internal membranes.

3.4.2 Autotrophic eukaryotes

But what of the autotrophic eukaryotes? How did they acquire their chloroplasts? We saw earlier that the thylakoids of cyanobacteria are very similar to those of the chloroplasts of land plants. This is no accident, and the similarity has lent support to the once-ridiculed idea that the chloroplasts of eukaryotic plant cells represent prokaryotic photosynthetic organisms like cyanobacteria which were engulfed by bigger, non-photosynthetic cells early on in evolution.

The evidence for this is now overwhelming. Molecular studies of chloroplast and cyanobacterial genes suggests that *all* chloroplasts must have been derived directly or indirectly from cyanobacteria. The only piece of evidence that contradicts this theory is that the chloroplasts of land plants contain the additional photosynthetic pigment chlorophyll *b*. Recently, however, three genera of photosynthetic bacteria have been discovered, the **Prochlorophyta**, which more closely resemble chloroplasts. They contain the pigments found in the chloroplasts of land plants—chlorophylls *a* and *b*, and carotenoids. Their thylakoids are also similarly packed together in distinct stacks (those of cyanobacteria are evenly spaced). This construction and pigmentation is that of the hypothetical ancestor of land plant chloroplasts (although molecular studies show that the Prochlorophyta are not closely related to the chloroplasts of present-day organisms).

The molecular evidence suggests that chloroplasts were first captured at around the time of the main divergence of the living eukaryotes, around 1–0.8 billion years ago. But, as we shall see in the rest of this book, many different groups of organisms seem to have since obtained chloroplasts, and hence the

ability to photosynthesize, by endosymbiosis. Not all appear to have engulfed cyanobacteria. In many instances organisms may have obtained chloroplasts by the more complex process of **secondary endosymbiosis**. Other photosynthetic eukaryotes seem to have been the starting point, which have subsequently been purged of their nuclear DNA. Chloroplasts thought to have been obtained in this way are surrounded by three or four membranes rather than the two seen in simple chloroplasts. The outer two layers are thought to represent the outer membrane of the engulfed organism and the membrane used by the host to engulf it.

Whether they engulfed cyanobacteria or other eukaryotes, however, endosymbiosis is probably the most important and commonest example in plant evolution of a convergent trend. Under similar selection pressure, eukaryotes seem to have repeatedly turned to 'stealing' to acquire photosynthetic machinery.

3.5 RADIATION OF THE EUKARYOTIC ALGAE

Since their emergence, the eukaryotes have undergone dramatic diversification. Within the groups of organisms that have become autotrophic, there have been two major evolutionary trends. Some have remained planktonic organisms and are mostly single-celled. The remainder have adopted a benthic lifestyle and become multicellular. However, it must be stressed that this functional division does not always reflect a taxonomic division. The history of the autotrophic eukaryotes has been marked by widespread divergence and convergence. Members of totally unrelated groups, such as red and brown algae, are often superficially very similar, while a single taxonomic group such as the green algae can contain planktonic and benthic forms which are very different. Endosymbiosis seems to have been a common occurrence and some groups, such as the dinoflagellates, even contain both heterotrophic and autotrophic organisms.

For these reasons we have made no attempt to introduce the eukaryotic **algae** in a strictly taxonomic order. The term alga does not, in any case, denote a strict taxonomic grouping but a functional one. They are 'aquatic plants', but a more precise explanation is impossible. There are, of course, aquatic plants that are not algae (see Chapter 11). Most of the algae are pretty simple (but not the giant kelps, see Chapter 4), most have no vascular or conducting tissue and most have simple reproduction (but not some red algae, whose life cycle is more complex than any flowering plant). We will avoid trying to divide this fascinating assemblage and will instead use the functional divide, examining the most important groups of planktonic algae in the second half of this chapter and introducing benthic forms in Chapter 4. There is space to cover only the most abundant and best-understood algae in this book and it is important to be aware that the following is not an attempt to give a comprehensive account of this enormous and diverse group of organisms.

3.6 THE DINOFLAGELLATES (PYRROPHYTA)

The Pyrrophyta, or **dinoflagellates**, have an incredible range of living conditions. Many are free-living autotrophs containing chloroplasts which they seem to have acquired from a range of sources. Others have no chloroplasts and are entirely heterotrophic, ingesting solid food particles like other flagellates and bacteria. Still others have themselves been the subjects of endosymbiotic events and are found, as much-modified **zooxanthellae**, within various animals such as sea anemones, jellyfish, giant clams and, most importantly, corals (see Chapter 4).

3.6.1 Structure

The main feature uniting members of this group is their mode of nuclear division which is sometimes described as mesokaryotic. Their chromosomes are always condensed (Fig. 3.5b), and have far less protein associated with their DNA than other eukaryotes.

Photosynthetic dinoflagellates have chloroplasts that contain chlorophylls *a* and *c*, and carotenoids including peridinin and fucoxanthin. A preponderance of the latter renders the cells of some dark

brown in colour. Many free-living members are protected, at least from small predators, by armour. The cell walls of these organisms are invested with thick plates of cellulose (Fig. 3.5) forming grooves along which run two powerful flagella. The flagellum which projects out at the back of the cell provides direction, and the other, which runs around the middle of the cell body, provides forward thrust which incorporates spin (Fig. 3.5a, b). These cells therefore cut through the water in a trajectory very similar to that followed by a bullet from a gun. Such cells have no trouble keeping themselves in optimum light conditions, travelling along gradients of nutrient, or escaping from hostile conditions or organisms.

Flagella clearly give flagellates a big advantage over non-motile cells in an aquatic situation, but their origins are far less clear than those of chloroplasts. Support for an endosymbiotic origin has waxed and waned, and the jury is still out. The evidence for endosymbiosis is in the form of the association of semi-autonomous DNA with basal bodies of flagella in the present-day green alga *Chlamydomonas*. The existence of a circularmapping genome, albeit far smaller than found in free-living bacteria, in chloroplasts and mitochondria is taken to represent part of the DNA of the original cell. Chloroplasts and mitochondria are autonomous, to the extent that they can replicate separately from the nucleus of the cell, but both depend heavily on the nuclear genome for the synthesis of molecules vital for both structure and function. Although there are no present-day bacteria with microtubules—an essential element of flagella—supporters of an endosymbiotic origin for flagella invoke a hypothetical prokaryotic ancestor with microtubules.

3.6.2 Ecology

Planktonic dinoflagellates are usually present at low densities, but some species are famous for production of the phenomenon of 'red tides'. These blooms of algae have much in common with cyanobacterial blooms but can be even more toxic. The cells of many of the species contain neurotoxins. These are not much of a problem for simple creatures such as shellfish, but since the shellfish concentrate the toxins, they are a major problem for humans or other

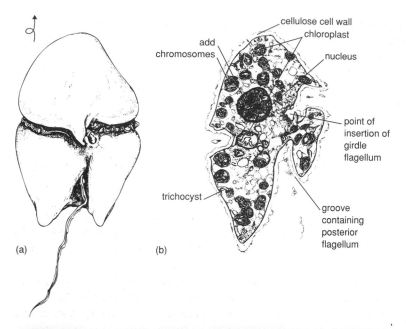

Fig. 3.5 Dinoflagellates. (a) A surface view (×2000) of a *Glenodinium sanguineum*, an organism responsible for red lakes. The girdle flagellum is much coiled, so only small sections appear in (b), a drawing from a transmission electron micrograph of the same cell. It shows the permanently condensed condition of the chromosomes in the membrane-bound nucleus, and other organelles, including **trichocysts** (complicated structures with no known function). The arrow shows the direction and bullet-like trajectory of this organism. (c) Scanning electron micrographs (×6000) of *Gonyaulax* and *Protoperidinium* (d) (flagella detach during processing for electron microscopy).

vertebrates that eat the shellfish. The toxins and their effects (the most acute 'shellfish poisoning' can lead to paralysis and death) are well characterized, but the causes of red tides are not. As with cyanobacterial blooms, nutrient-rich run-off or sewage from the land are thought to play a part. Members of one dinoflagellate species, *Pfiesteria piscicidia*, actually kill fish with their (extracellular) toxin, and the cells then feed upon the fish as it decays.

Another curious attribute of some dinoflagellates is their ability to glow in the dark. Using the same mechanism as fireflies, dinoflagellates have two sorts of luminescence. One occurs in a regular diurnal rhythm and its function is unknown. The other occurs in response to physical stimulation such as wave-break, oars passing through the water, footfalls on a beach, or even submerged submarines. Aircraft can make use of this response to detect the submarines, or for more peaceful purposes such as spotting shoals of fish for fishing fleets. The fact that the dinoflagellates respond so strongly to fish may indicate the significance of the bright response to stimulation; it may act as a deterrent to predators.

3.7 THE DIATOMS (BACILLARIOPHYTA)

The Bacillariophyta, or **diatoms**, relatives of the brown algae we shall examine in the next chapter, are another group of organisms with great importance in terms of planktonic productivity. It has been estimated that planktonic diatoms account for around 25% of the world's primary production. They are also amongst the most abundant planktonic organisms in an enormous diversity of habitats ranging from fresh water to marine environments. A typical litre of sea water contains several million of these tiny algae; water from a bloom, the formation of which is a regular feature of waters in Springtime in temperate zones, contains even more.

3.7.1 Structure

These algae share several of the survival tricks used by organisms described above. Virtually all are autotrophic, containing chlorophylls *a* and *c*, green pigments which are often completely masked, as in brown algae, by the brown carotenoid fucoxanthin. Some, however, can switch between autotrophy and heterotrophy, using organic substrates, such as in sediments, rather like some cyanobacteria. Some diatoms are obligate heterotrophs, and parasitize other plants or animals.

Diatoms are armoured, like the dinoflagellates, but their walls, or **frustules**, are strengthened with silica rather than cellulose. These rigid-walled species are delightfully elaborate (Fig. 3.6), and are of two basic forms, both similar in basic structure to a tightly lidded Petri dish. One body form is essentially radially symmetrical (**centric**, Fig. 3.6a), the other bilaterally so (**pennate**, Fig. 3.6b).

Neither type of diatom has flagella except when centric diatoms generate male gametes. Many pennate species are capable of locomotion, however, and they have an elegant, albeit slow, gliding motion. Exactly how the movement is effected is not understood, but the most favoured theory at present has much in common with the movement of a military tank on its tracks. It is thought that the cells release polysaccharide material through a slit or **raphe** (Fig. 3.6b) which runs around the cells. Fibrils of polysaccharide adhere to the substrate outside the diatom at one end, and to the outer membrane of the diatom at the other. The membrane is moved under the raphe, possibly by microfilaments of actin, as in human muscles and in cyanobacteria. This pulls the diatom over the substrate in the opposite direction, just like the treads of a tank. The ability to move is especially important to the many diatoms that live in mud or sand, as they can move up through the substrate when it is light or wet and migrate down when it gets too dry for them.

3.7.2 Cell division

Diatoms are truly eukaryotic, unlike the dinoflagellates, but those with the most rigid frustules face quite a problem when it comes to cell division. They remain protected from the exterior environment by using the 'lid' and 'base' of the original cell to make two new bases. That is all well and good for the

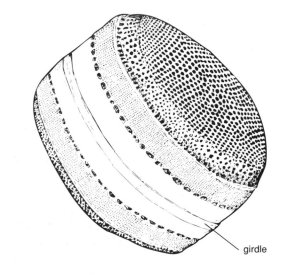

(a)

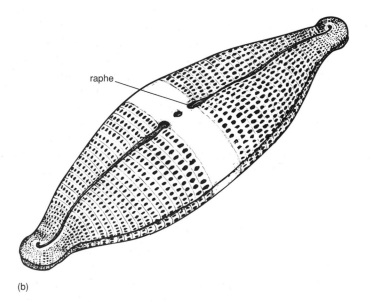

(b)

Fig. 3.6 Diatoms as drawn from scanning electron micrographs of dead specimens to show the elaborate structure of the cell walls. (a) A centric diatom (×800). The girdle is the junction between the 'lid' and 'base' of the cell covering. (b) A pennate diatom (×800). In life there would be material protruding out of the raphe canal.

daughter formed from the lid—which turns out the same size as the parent cell—but the other daughter is slightly smaller. Over time the average size of cells in populations of rigid-frustuled diatoms therefore steadily declines. The solution to the problem is to produce spores or undertake sexual reproduction. As these processes are far better understood in the green algae than in any other group we will leave their consideration until we have introduced this important assemblage.

3.8 THE UNICELLULAR GREEN ALGAE (CHLOROPHYTA)

3.8.1 Structure

We used the word assemblage for these algae, because it is difficult to argue that they form a coherent group. They range from tiny non-motile unicells, through motile unicells and colonies, to macroscopic plate-like, tubular, branched and unbranched forms.

What unites the chlorophytes is the possession of chloroplasts (obtained by endosymbiosis of cyanobacteria) which contain chlorophyll *a* and *b*, with carotenoids as accessory pigments. They are true eukaryotes and most have cells surrounded by a strong cell wall composed of cellulose fibres embedded in a hemicellulose matrix. Together with their cellular ultrastructure, this makes them more like terrestrial plants than any other group of algae. Indeed, it is members of this assemblage that are thought to have generated the ancestors of the terrestrial plants, as we shall see in Chapters 4 and 5.

3.8.2 Ecology

Most unicellular chlorophytes inhabit predominantly wet environments, but their diversity is reflected in the enormous range of habitats in which they live: they thrive in the tiny water spaces between soil particles; in rivers and lakes; and even in the saltiest of the seas, the Dead Sea. The organisms capable of tolerating these very different environments have a variety of structural and physiological adaptations that account for their success. Freshwater forms have contractile vacuoles which act like tiny pumps to remove the excess water which moves into their bodies by osmosis. In contrast marine forms produce unusual end-products of metabolism such as glycerol, which increase their internal osmotic potential and stop them losing water.

But although most are planktonic organisms, some unicellular chlorophytes can live away from bodies of water. Some of these live in the frigid cold of the Antarctic, routinely inhabiting snow (see Chapter 10); others are even capable of surviving in terrestrial environments. The latter add the bright-green tinge to the Northerly facing aspects of tree bark and monuments you might have noticed in all but the driest of climates. These algae have phenomenally hygroscopic (water-absorbing) cell coverings, which can suck in vast amounts of moisture—up to 20 times their dry weight—really fast when it is available.

As no one species could therefore be thought of as 'typical' of the group, we have picked the species that has been subject to the most intense research to illustrate this versatile and highly successful assemblage of organisms.

3.8.3 *Chlamydomonas*

The tiny biflagellate cells of this genus (Figs 3.7 and 3.8) live in almost every place on earth, and indeed they are of great environmental significance, but the easiest place to find them is in a research laboratory! The reasons for this are simple. They are easy to 'cultivate', being undemanding in their nutritional requirements and tolerant of a wide range of temperatures, densities and light levels. They divide and grow quickly, and their cell cycle can be synchronized to a light–dark cycle (Fig. 3.7b) so each event from inception of a new cell to formation of the next can be predicted and studied with confidence. Their usual state is haploid, so mutations are relatively easy to spot, and a very wide range of mutants have been isolated. These mutants have allowed us to find out what controls expression of an enormous range of processes essential to all life forms including photosynthesis, cell division, recognition and mating. Perhaps the greatest contribution *Chlamydomonas* has made to our understanding is in the latter two of those processes, and as the development of sexual systems was such an important evolutionary step, we will now consider reproduction in this genus. It is important to remember the great diversity within the chlorophytes, however, as there are many other very different methods of both vegetative and sexual reproduction practised by other genera.

3.8.4 Asexual reproduction

The division of one cell of *Chlamydomonas* into two is a process which superficially resembles that in cyanobacteria more closely than that of other green plants. Cells become 'pinched off' by the growth of cell-wall material *in* from the outside of the parental cell, just like those in the unicellular blue–greens. The wall material that separates the two daughters of most green plant cells, in contrast, starts to grow from the *inside* of the parental cell, and grows to meet the exterior, and in this respect *Chlamydomonas* can therefore considered to be rather primitive. As we can see in Fig. 3.7b (p. 48) this process can occur twice in 24 h, and if the cells are synchronized to a light–dark cycle both divisions will occur at night. It is the dawn (light) which triggers

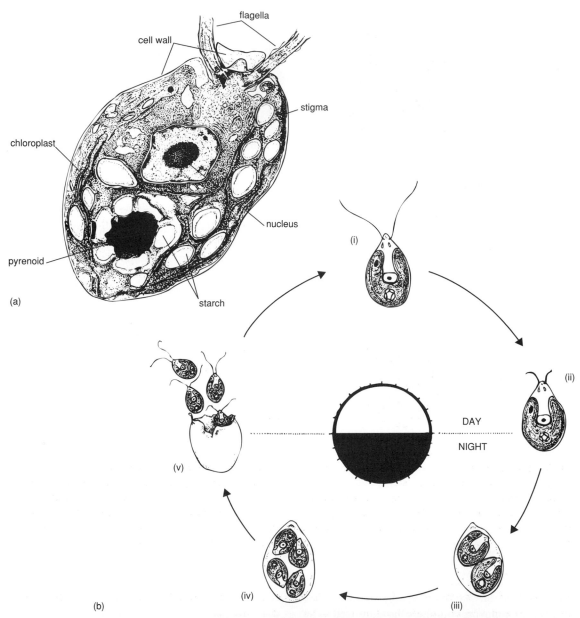

Fig. 3.7 (a) Drawing (×700) from a transmission electron micrograph of *Chlamydomonas reinhardtii* showing the single, basin-shaped chloroplast and the 'stigma'. The latter is sometimes called an 'eyespot' because it is the site concerned with light (photo-) reception and it appears as a characteristic red spot under the light microscope. The redness comes from rhodopsin (as in many animal light-perceptive structures). The two flagella are truncated in this figure, as they would be too long to fit on such a drawing, therefore see (b) for their general appearance. When beating, they pull the cell through the water in a 'breast-stroke in a spin' manner. (b) Stylized representation of asexual reproduction in *Chlamydomonas*. In a laboratory environment with suitable temperature and lighting control, this cycle would happen once every 24 h. (i) The 'day' form: a motile, photosynthetic unicell. As night falls (or the timer switches the laboratory lighting off) the flagella regress (ii) and one mitotic cell division occurs (iii), within the protection of the original cell wall. Another division (iv) produces a total of four cells from the original one, which grow their own flagella and are released (v) when light hits the original cell wall.

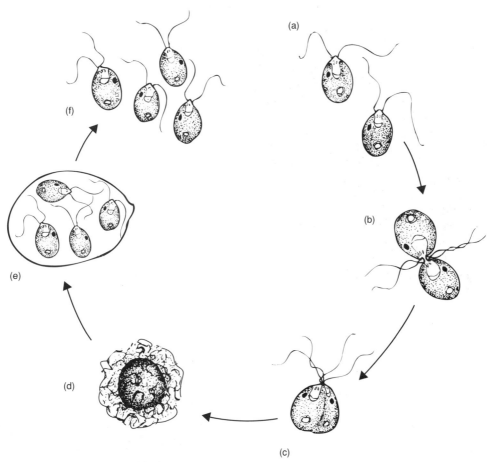

Fig. 3.8 Stylized representation of sexual reproduction in *Chlamydomonas*, which exhibits **zygotic meiosis**. (a) A pair of haploid cells which look identical but which are two mating types (plus and minus). (b) Agglutination along the flagella has brought the two cells into intimate contact and they have started to fuse. At this point the flagella have stopped making glycoproteins, so have started to detach from each other. (c) The fused cells may swim about for some time in this form. (d) Especially in harsh conditions, the newly formed zygote secretes a thick wall and may remain dormant for some time before meiosis occurs, forming four haploid progeny (two of the plus mating type, two minus) (e) and (f).

the release of an enzyme which breaks down the mother cell wall, whereupon the four new cells break out and swim off, ready for a day of growth and photosynthesis. As darkness falls, their flagella regress, they are disassembled and the building blocks stowed away, ready for recycling into the daughter cells, and the process starts all over again. With sufficient light, the right temperatures, and adequate minerals, this cycle can carry on endlessly, but two common hazards may prevent this. They are desiccation—if the soil or pond dries out—and min-

eral depletion, which would leave too low a level of essential salts.

Dehydration causes *Chlamydomonas* to forsake its usually vigorous breast-stroke-like swimming motion and develop a mucilaginous exterior. This non-motile palmelloid stage can survive relatively dry spells. The return of water prompts the re-emergence of the flagella and the resumption of activity. Mineral depletion, in particular nitrogen, leads to a more extreme response.

3.8.5 Sexual reproduction

When mineral concentrations fall below a certain level, the simple cycle outlined above ceases, and the cells invest their energy in becoming gametes rather than in dividing. In the simplest species there is nothing to betray their new status to the human eye, but each *Chlamydomonas* cell is now either one 'sex' or the other. The two types are therefore referred to as plus and minus, rather than male and female. It would be wonderful if we knew just how such differences first arose in evolution, but the study of *Chlamydomonas* can at least reveal a wealth of information about how different mating types manage to recognize, meet and interact with each other. The beauty of knowing the trigger for gamete production in *Chlamydomonas* is that by transferring a laboratory culture into a low nutrient medium, we can readily observe and study these events.

The most important difference between vegetative cells and gametes is in their surface molecules. Gametes have special glycoproteins, or **agglutinins** on their flagella. These are inherently sticky molecules which cause sexually competent individuals to stick to each other. The agglutinins are slightly different on plus and minus cells, and when compatible cells meet, there is contact activation. This means that where a flagellum of a plus cell touches that of a minus cell, more agglutinins are produced and the stickiness is enhanced, until the cells have become glued together all along their flagella. This brings the bodies of the cells into contact, and fusion of the cells soon follows (Fig. 3.8b). The mating process therefore owes much to surface recognition molecules in *Chlamydomonas*, and we now know that glycoproteins are important in a very wide range of signalling processes in both plants and animals.

There are other differences between gametes and vegetative cells of even the simplest *Chlamydomonas* species, notably in the reserves held by the cells, which will form the basis of the reserves needed by the zygote which forms when cell fusion is complete. Some *Chlamydomonas* go one step further, however, and the gametes even look different. Such species perhaps give a clue about evolutionary progress, which seems to have reflected increasing metabolic investment in the parent destined to become the 'mother' and decreasing investment in the other. In some species, the gametes are similar but of different sizes. The slightly larger gamete is the female parent, which is packed with reserves lacking in the male. In what are assumed to be more advanced species, the female parent is still larger and non-motile. Only the males are motile, and this resembles the situation in many of the most sophisticated algae, bryophytes and ferns. One crucial extra strategy of such plants is lacking in *Chlamydomonas*, however; there are no long distance signalling molecules, and gametes must literally bump into each other. The development of **pheromones**, molecules capable of allowing recognition at a distance, was an evolutionary step reserved for other multicellular groups such as the brown algae we will meet in the next chapter.

Whatever the structure of the gametes, however, the events following on from gamete fusion in *Chlamydomonas* are similarly well choreographed for survival of hardship. The now diploid zygote, after a swim (which may help dispersal), loses its flagella and secretes a highly resistant special wall. This wall serves the same purpose as those of the resistant spores and akinetes of cyanobacteria, protecting the organism from dehydration and inclement conditions. After a period of dormancy, when adequate water or permissive mineral levels are restored, the zygote undergoes meiosis and the haploid condition is restored. The four products of the division develop flagella and swim off to begin the asexual cycle once more.

The evolutionary steps that led to sexual reproduction were clearly popular, as most of the benthic algae and the remainder of the plants we will consider in this book indulge in it at some point in their life cycles. The potential benefits conferred by the mixing of two genomes were discussed in detail in Chapter 1, but it is important to note that there are many very successful algae and plants that never bother.

3.9 POINTS FOR DISCUSSION

1 What advantages do you think the internal compartmentation of eukaryotes confers?

2 Why are stromatolites only found in salty environments, whereas early species were widely distributed?
3 What might be the adaptive advantage to dinoflagellates of producing neurotoxins?
4 Why are multicellular photosynthetic organisms so rare in the plankton? Why are the majority of multicellular floating forms filamentous?
5 What do you think are the disadvantages of sex for *Chlamydomonas*?

FURTHER READING

Bhattacharya, D. & Medlin, L. (1998) Algal phylogeny and the origin of land plants. *Plant Physiology* **116**, 9–15.

Bold, H.C. & Wynne, M.J. (1978) *Introduction to the Algae*, 2nd edn. Prentice Hall, New Jersey.

Copley, J. & Graham-Rowe, D. (1999) The cold war resurfaces [bioluminescence in algae and submarine detection]. *New Scientist* **2213**, 4.

Knight, J. (1999) Blazing a trail [bioluminescence in dinoflagellates]. *New Scientist* **2213**, 28–32.

Lee, R.E. (1999) *Phycology*, 3rd edn. Cambridge University Press, Cambridge.

Raven, P.H., Evert, R.F. & Eichorn, S.E. (1999) *Biology of Plants*, 6th edn. W.H. Freeman, New York.

CHAPTER 4

Life on the rocks

4.1 INTRODUCTION

There could not be a stronger contrast between the planktonic algae we examined in the last chapter, which float free in open water, and the benthic algae, which live attached to the rocky beds of rivers, lakes and sea shores. The planktonic algae are mostly small single-celled organisms, which, like many unicellular heterotrophs, are often highly motile. In contrast the benthic algae such as the seaweeds are much larger and more complex, often showing division of labour between different types of cell and between different parts of their bodies. They are altogether much more like land plants. This chapter is in three main parts. First, it examines the fundamental reasons for the major differences between planktonic and benthic algae. Second, it describes something of the diversity of the form, life history and ecology of the groups of organisms that have adapted to a benthic way of life: the green, red and brown algae; and the corals. Finally, it investigates why such a diversity of these benthic organisms continues to survive.

4.2 SELECTIVE FORCES AND EVOLUTION OF BENTHIC ORGANISMS

Benthic algae probably evolved early in the history of the eukaryotes, but we know little about their early evolution because their soft bodies do not fossilize well. However, stages in their evolution can be reconstructed by considering the selective forces that drive this process.

4.2.1 Growth and form

We saw in the last chapter that the needs of **planktonic** algae for adequate light and for a supply of carbon dioxide and nutrients has had a great influence on their design. Their small size, neutral buoyancy and motility allow them to stay up near the surface of the deep oceans; and their small size also means that diffusion can supply them with adequate resources. Selection for unicellularity will be strong. However, the requirement for light and nutrients can lead to quite different selection pressures on the organisms that live in shallow water.

Advantages of multicellularity in benthic algae

Light can penetrate right to the bottom in shallow water, and even organisms that sink to the floor can photosynthesize. The way is therefore open for **benthic** algae to evolve. Indeed, if they attached to the bottom, algae could actually obtain some advantage over the floating forms. Water currents, which will flow past them, would supply them with nutrients at a faster rate than diffusion can supply cells that are floating with the current. Benthic organisms therefore would be able to grow faster, so once they were in this advantageous position they should stay put. Rather than reproducing sexually and releasing motile spores it would pay them to divide asexually. In this way they would both maintain their position and colonize the adjacent areas where conditions would also be favourable. There should therefore be strong selection pressure on benthic algae to evolve multicellularity.

Advantages and disadvantages of horizontal growth

The directions in which the cells divide, and therefore the shape of the organisms that they produce would also influence their chances of survival. One obvious solution is for cells to divide in the horizontal plane to produce a single layer of **encrusting** cells covering the surface (Fig. 4.1a, b). This solution has the advantage that the organism prevents neighbours from attaching while maintaining a reasonable surface area for photosynthesis, but it suffers from several disadvantages. First, the flow rate of any body of water is reduced towards the bottom, falling to zero at the interface between the water and the solid floor. There is therefore effectively a 'boundary layer' of slow-moving water above the bed of any water channel through which gases and nutrients must diffuse to reach the cells. The growth rates of encrusting organisms are consequently limited. Second, encrusting organisms take up a great deal of area for their size, so their growth will soon be stopped when they abut against their neighbours. Third, encrusting algae would be outcompeted by algae that project above the surface and cast them into shade. For these reasons encrusting algae are rare except in very rough areas or where light levels are very low. Most benthic algae instead grow away from the surface, to produce a deeper bed of vegetation.

Advantages and disadvantages of vertical growth

The simplest way for an alga to grow away from the surface is for its cells to divide in the vertical plane. This produces a one-dimensional **filament** (Fig. 4.1c) which projects out into the flow and towards the light. A branched filament (Fig. 4.1d) would be an even better competitor for light, because it would display a larger area and cast a deeper shade, just like a tree. However, even this design will be inferior to a two-dimensional plate of cells or **thallus** (Fig. 4.1e).

Although all of these structures compete more effectively for light than encrusting forms, they also inevitably suffer from new problems. Because their bodies project further out into the flow they are subjected to greater hydrodynamic forces which tend to pull them off their rocks or tear them apart. The lower regions of the organisms are also overshadowed and sheltered by the upper regions, so are not able to provide useful amounts of photosynthesis. As a result there is strong selection pressure for division of labour in these organisms. The upper regions need to concentrate on photosynthesis and extension

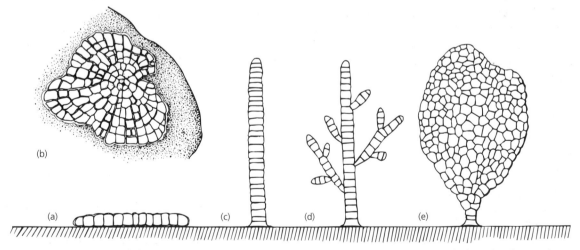

Fig. 4.1 Alternative growth forms in multicellular benthic algae. **Encrusting** forms (a) are deep in the boundary layer and so will be sheltered from excessive currents. A good example is the green alga *Coleochaete* (b), seen here in plan view (×60), which is a close relative of land plants. However, encrusting algae will have slow growth rates and can be shaded out by forms that grow away from the substrate. More common forms in relatively slow-moving water are **filaments** (c), branched filaments (d) and plate-like **thalli** (e) which project out into the flow and are increasingly good competitors for light.

growth. In contrast the lower regions, particularly the attachment or **holdfast** between the organism and the rock, must be strengthened by mechanical structures. Regions in between must be modified into transport structures that move sugars down from the top to the bottom. Because there is a limit to the strength and transport efficiency of single cells, very large organisms have to be several cell layers thick, especially near the bottom. The outer cells can be specialized for photosynthesis, while the inner ones, which are shaded by the outer cells, are specialized to perform the mechanical and transport functions. As we shall see, algae have evolved several different ways of producing such a body. Whichever way they do it, however, each cell in the body is specialized and will be dependent for its survival on all the others. Benthic algae are therefore the first examples we shall see of truly integrated multicellular organisms.

4.2.2 Reproduction and life history

If an organism is to do well it must not only survive and grow but also colonize new areas before it dies. A large alga could theoretically reproduce asexually by splitting itself up into smaller parts from time to time and allowing itself to float to new habitats. However, this strategy has several disadvantages. It would be hard for a large complex organism to reattach successfully at its holdfast region. Furthermore, floating organisms would no longer get water flowing past them, so their supply of carbon dioxide and nutrients would be drastically reduced. Parts of algae that broke off the parent would therefore be likely to die.

A much better option is to reproduce sexually, releasing small spores and gametes which are capable of surviving afloat and will be dispersed by water currents. Finally the spores, or the zygotes formed by fusion of the gametes, can settle and reattach to the substrate. Therefore it is no surprise that the spores and gametes of multicellular algae strongly resemble the planktonic algae which we examined in the last chapter; all are unicellular and metabolically active and many have flagella. Indeed they often closely resemble the unicellular algae from which the adult organisms evolved.

4.3 GROUPS OF MULTICELLULAR ALGAE

Since they evolved in response to the same selection pressures it is no surprise that all three major groups of multicellular algae show a certain degree of convergent evolution: green, brown and red seaweeds can be remarkably similar in form.

However, the groups do differ in complexity, form and in the habitats they characteristically occupy, in ways that can be related to the photosynthetic pigments and to the other aspects of cell structure that they inherited from their unicellular ancestors. Their life histories also show a great deal of diversity, much of which derives from their evolutionary heritage.

4.3.1 The green algae (Chlorophyta)

As we saw in the last chapter, the Chlorophyta have grass-green chloroplasts which contain the photosynthetic pigments chlorophyll *a* and *b* and β carotene, and preferentially absorb both red and blue light (Fig. 4.2). This suits them only to life in shallow water, because these wavelengths are preferentially filtered out by water. The Chlorophyta are therefore particularly common in fresh water where they are by

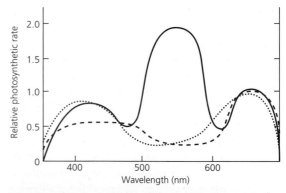

Fig. 4.2 The effect of the wavelength of illuminating light on the photosynthetic rates of green (dotted line), red (solid line) and brown (dashed line) algae. Red algae preferentially absorb the green–blue light which can penetrate water, so thrive in deeper water. In contrast green algae and, to a lesser extent, brown algae absorb light more at the red and blue extremes of the spectrum, and so are better suited to shallow water.

far the most successful of these groups of multicellular algae, and they demonstrate great diversity, both in their structure and their life histories. However, despite their relationship to the land plants, and their strong cellulose cell walls, the Chlorophyta themselves tend to be small, fragile organisms living in sheltered habitats. Multicellularity seems to have developed several times within the group but in no case do they show such sophistication in design as the red or brown algae.

The Ulvophyceae

Of the three major classes of the Chlorophyta, the Ulvophyceae, which are primarily marine, include a range of forms which demonstrate interesting developments in multicellularity. The many filamentous algae that are found in this group include members of the genus *Cladophora* which grow in dense mats in slow-moving estuaries or rivers and which may be attached to the bottom or free-floating. Another species is *Ulothrix zonata* which attaches to stones in its freshwater habitat via a holdfast cell. A slightly more complex organism than these filamentous algae is the sea lettuce *Ulva* (Fig. 4.3a) which has a plate-like thallus that is two cells thick. It grows on sheltered shores and in rock pools where it is attached to the rocks by a multicellular holdfast which is reinforced by long 'hyphae' which extend from basal cells into the rock below (Fig. 4.3b, c).

The other organisms in this class are the **coenocytic** algae which possess a rather strange mix of unicellular and multicellular characteristics. At first glance these appear to be large multicellular organisms. However, closer examination reveals that they consist of just a single huge cell containing multiple nuclei. During their growth, their nuclei repeatedly divide without any cell walls being formed to divide up the alga. This pattern of development is similar to that seen in the mycelia of advanced fungi. Because of the lack of cell walls to brace the structure, these algae are rather fragile and most inhabit sheltered tropical waters. *Ventricaria* simply resembles a bladder but the many species of *Caulerpa*, the so-called 'sea grapes', show striking convergence to land plants (Fig. 4.4a–f). They develop organs analogous to stems, leaves and even roots which anchor them in

the sand of the sheltered coral lagoons they tend to inhabit. One species *Caulerpa taxifolia* has recently escaped into the Mediterranean and is spreading fast, threatening to outcompete all other seaweeds and devastate the ecology of the region. A few coenocytic algae, such as *Halimeda* (Fig. 4.4g), have been able to colonize quite rough waters by calcifying their cell wall and so strengthening it. The soft white sand characteristic of tropical beaches is formed by the breakdown of this alga which dominates many coral reefs throughout the tropics.

The Chlorophyceae

The second class of the green algae, the Chlorophyceae, contains mainly single-celled or colonial organisms, some of which we examined in the last chapter. However, it also contains a few filamentous algae, like *Stigeoclonium*. Some are even more complex. *Fritschiella* has a rhizoid, a thick lower body and two types of erect branches, which adapt it to a terrestrial existence on damp surfaces. In many ways it probably shows convergent evolution with the early land plants that we will examine in Chapter 6.

The Charophyceae

The third class of green algae, the Charophyceae, is the group that is most closely related to the land plants. As well as having the most similar genetic make-up, they also share a similar pattern of cell division. During this process the two nuclei are separated by a central structure called a **phragmoplast**, rather than by ingrowth of the outer walls —a **phycoplast**—which is typically found in the Chlorophyceae.

The Charophyceae include the filamentous freshwater alga *Spirogyra* which is beloved of school teachers. But probably the most important members of the group, because the closest living relatives of the land plants, are the stoneworts, order Charales. Stoneworts are found in brackish and fresh water and are so called because they produce calcified cell walls. *Coleochaete* (see Fig. 4.1b) is a simple plate of cells, like the simplest forms of land plants. *Chara* (Fig. 4.5) exhibits a rather more complex structure which

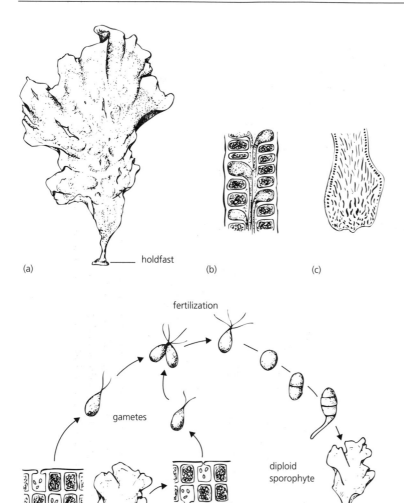

(a) holdfast (b) (c)

(d)

fertilization

gametes

diploid
sporophyte

meiosis

meiosis

spores

haploid
gametophyte

Fig. 4.3 The thalloid green alga *Ulva*, seen (×0.3) in plan view (a). A longitudinal section of the lower thallus (b) shows the hyphae which grow down between the two rows of cells to the holdfast, reinforcing lower regions. A longitudinal section of the holdfast region (c) shows the build-up of many layers of hyphae. The life cycle of *Ulva* (d), shows that it undergoes **sporic meiosis** with equal development of the haploid and diploid generations. Spores and gametes are produced by individual surface cells on the thallus of diploid and haploid adults, respectively: both are motile, but spores have four flagella whereas gametes have only two.

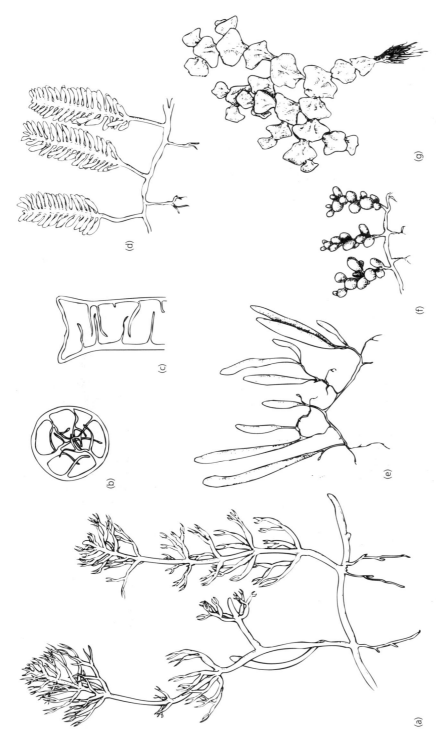

Fig. 4.4 Coenocytic algae. The bodies of these algae are unicellular but multinucleate. Like many other species *Caulerpa verticillata* (a) (×0.3) superficially resembles multicellular seaweeds and land plants. However, study of transverse (b) and longitudinal sections (c) of its stem shows the lack of dividing walls. The structure is instead strengthened by a mass of internal bracing struts. Other species of *Caulerpa* include *C. floridana* (d), *C. prolifera* (e), and *C. racemosa* (f), (all ×0.25) which possesses the bulbous 'leaves' which give the name 'sea grapes'. *Halimeda opuntia* (g) (×1.0) has a cell wall strengthened by impregnation with calcium carbonate.

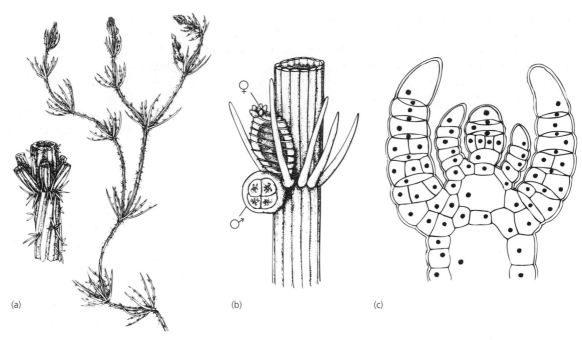

Fig. 4.5 Stoneworts, close relatives of land plants. (a) Plan view of *Chara hispida* (×1.0), showing the whorls of branches radiating from a long central axis. (b) *C. vulgaris* (×4.0), showing a branchlet node with reproductive **gametangia**. (c) Longitudinal section of the stem apex showing the **corticated filament** structure with large central cells, used for support and transport, surrounded by a layer of small photosynthetic cells.

may help it survive in faster-moving water. Its cells divide in such a way as to produce several whorls of branches (Fig. 4.5a), each of which radiates from nodes along a central stem-like axis. The axis itself also has complex organization, showing a division of labour. It is a **corticated filament** (Fig. 4.5c), with large central cells which are surrounded by a sheath of smaller ones. The two sorts of cells provide complementary functions. The small outer cells photosynthesize, while the thick walls and wide lumen of the central cells provide mechanical support and allow sugars to be transported down the organism by cytoplasmic streaming. *Chara* resembles primitive land plants in its possession of branched **rhizoids**, which help anchor it in mud. As we shall see this development was probably crucial in the invasion of dry land.

4.3.2 The red algae (Rhodophyta)

The red colour of many of the Rhodophyta which

gives the group its name is produced by their chloroplasts, which were derived from the endosymbiosis of cyanobacteria. They contain not only chlorophyll *a* and *d* but also the accessory pigments phycobilins. These pigments preferentially absorb the green and blue–green light that penetrates furthest through water, so the red algae are particularly suited to life in deep water. As a result, perhaps, few red algae live in fresh water which is usually rather shallow. However, there is a great diversity of over 5000 marine species, many of which are quite large and complex organisms which dominate the lower subtidal shore over much of the world.

Cellular structure

One factor which undoubtedly helps red algae survive in the sea is the rubbery nature of their cell walls. These are made of a soft mucilaginous matrix, reinforced by strong cellulose fibres which wind around the cell like the coils of a spring. This gives the walls a combination of strength and flexibility which allows red algae like *Chondrus crispus* (Fig. 4.6a) to flex away from waves and currents without breaking. In contrast red algae like *Galaxura obturata* (Fig. 4.6b) incorporate calcium carbonate into their cell walls. This helps protect them from being eaten by herbi-vores, but, paradoxically, their greater rigidity makes them more vulnerable to being damaged by waves. These 'coralline algae' therefore mostly inhabit calm subtidal regions, where they account for much of the productivity of coral reefs. A compromise design, developed by species such as the temperate *Corallina*, has flexible joints between the rigid regions, which enable it to resist both predation and wave action.

Growth forms

There is a great diversity of structural design within the red algae. As in green algae there are filamentous forms, such as the freshwater *Batrachospermum*, and plate-like forms, such as *Porphyra*, whose edible fronds are just one cell thick and are the basis of a billion dollar seaweed farming industry in North-East Asia. However, some red algae are very different. There are several species of encrusting red algae that inhabit one of two very different habitats. Some have been found in the deep sea, at depths of up to 260 m, where there is not enough light for other more complex algae to grow. Others are common around the high tide mark where only encrusting forms can resist the desiccation and the severe action of the waves.

The majority of the larger and more conspicuous forms, though, are three-dimensional. They are formed by encasing a many-branched filamentous structure in a sheath of mucilage, rather like hair in a gel (Fig. 4.7a–d). This simple process of con-solidation is known as **pseudoparenchymatous** construction, and is also used by fungi to produce mushrooms and toadstools from consolidated hyphae. In the case of red algae, photosynthesis occurs in each of the terminal cells of the filamentous branches, which together make up what looks like its outer surface. The shaded inner cells, meanwhile, are strengthened for support. Pseudoparenchymatous structures are straightforward to construct because each cell only has to divide in one direction, just as in filamentous structures. However, cells in such a structure are only strongly attached to cells in their own filament. The join between adjacent filaments is weak, so pseudoparenchymatous structures will tend to be vulnerable to damage. There is also poor communication between the different parts of the

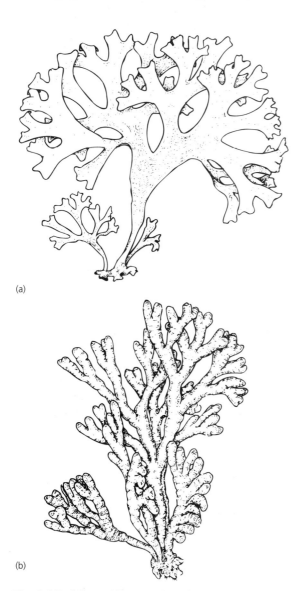

(a)

(b)

Fig. 4.6 Red algae with contrasting adaptations to resist wave action and predation. The walls of *Chondrus crispus* (a) (×1.0) have a soft mucilaginous matrix conferring flexibility, while those of *Galaxura obturata* (b) (×1.0) are hardened by calcium carbonate.

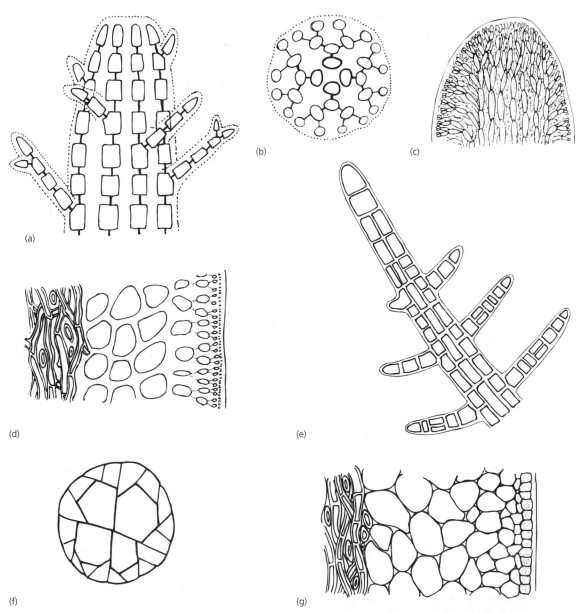

Fig. 4.7 Forms of three-dimensional construction. (a–d) **Pseudoparenchyma** in multiaxial Florideophyceae. (Redrawn from Seagel *et al.* (1982.) (a, b) Diagrammatic cellular detail of filaments in longitudinal (a) and transverse (b) cross-section. (c) Idealized apex in longitudinal section. (d) Longitudinal section of a real organism showing filamentous medulla and pseudoparenchymatous cortex. (e–g) **Parenchyma** in the Phaeophyta. (e, f) Longitudinal and transverse sections of the apical portion of *Sphacelaria* thallus. (g) Longitudinal section of *Halidrys*.

organism because any message can only travel along the filaments and not between them. To avoid these problems, therefore, some of the filaments of the larger red algae may be joined by secondary connections which form after cell division has ceased.

Pseudoparenchymatous red algae have proved extremely successful. The more conspicuous and well-known species include the 'Irish moss' *Chondrus crispus* (Fig. 4.6a), which is an important source of the paint additive, carrageenin; and the tough *Mastocarpus stellata*, which grows up to 20 cm in length and thrives on quite exposed shores.

4.3.3 The brown algae (Phaeophyta)

The chloroplasts of brown algae, like those of their unicellular relatives the diatoms, contain, along with chlorophyll *a* and *b* and carotenoids, the accessory pigment fucoxanthin. This preferentially absorbs red and blue light, but to a lesser extent than in green algae. These pigments give many of the algae a brown colour, and suit them for life in water up to 30 m deep. Although there are only 1500 species of brown algae, none of which grows in fresh water, they are the largest and most prominent seaweeds, dominating rocky shores outside the tropics. This dominance is due not to any special advantages of their photosynthetic apparatus or reproduction, but largely to the efficiency of their structural design.

Growth forms

The brown algae have strong, flexible cell walls very like those of red algae. The main reason for their success is, however, the form of construction they have developed. Although some of the smaller brown algae are filamentous or pseudoparenchymatous, the larger more complex types have evolved more efficient **parenchymatous** construction (Fig. 4.7e–g). In these organisms the meristematic cells divide in all three planes to produce a block-like three-dimensional structure just like the parenchyma of land plants. Because cells are firmly joined to all their neighbours by narrow tubes called **plasmodesmata** this structure is not only strong in all directions but allows excellent communication between different parts. This pattern of growth also enables organ-

isms to produce a complicated internal structure in which cells can be readily specialized for a range of different tasks (Fig. 4.7g). The small cells at the surface are specialized for photosynthesis, while many internal cells are elongated along the main axis of the thallus. Some have thickened cell walls and strengthen the thallus against tearing; others are modified and transport food. They have broad, open lumens with sieve plates at their ends just like the phloem cells of land plants (see Chapter 6). In large kelps, sugars and amino acids may be transported down the stipe through these cells at speeds of up to 60 cm/h, a rate similar to that in the phloem of land plants.

Groups of brown algae

The two major orders of brown algae are the Fucales and Laminariales. The Fucales, or wracks, are dominant along the intertidal zones of temperate shores, and include the seaweed which many of us remember from childhood holidays, the bladderwrack *Fucus vesiculosus* (Fig. 4.8). Its flattened fronds grow up to a metre in length and are held up to the light at high tide by the air-filled vesicles which burst with such a satisfying pop when squeezed. An even larger member of the group is the Caribbean species *Sargassum muticum* which grows up to 20 m in length. Plants of this species often break free from their moorings and can be found floating to the north-east in the huge 'rafts' characteristic of the Sargasso sea.

The Laminariales, or kelps, are even larger organisms than the Fucales, and have the greatest degree of differentiation into organs of any alga (Box 4.1). Their sophisticated design has allowed them to dominate the subtidal regions of temperate seas.

4.4 REPRODUCTION IN BENTHIC ALGAE

We saw in Chapter 1 that multicellular organisms that reproduce sexually can have one of three sorts of life cycles: the haploid stage could be the only one that becomes multicellular; the diploid stage could be the only one that becomes multicellular; or both haploid and diploid stages can become multicellular.

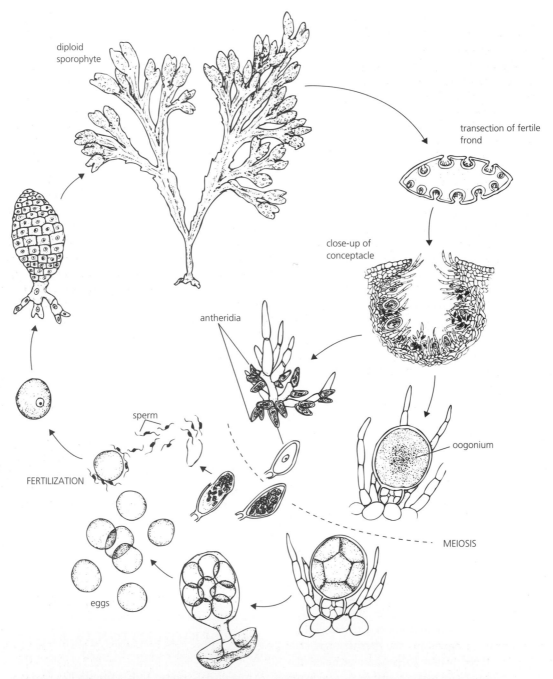

diploid
sporophyte

transection of fertile
frond

close-up of
conceptacle

antheridia

oogonium

sperm

MEIOSIS

FERTILIZATION

eggs

Fig. 4.8 The life cycle of the brown alga *Fucus vesiculosus* shows that it undergoes **gametic meiosis** with just a single diploid adult generation. The tips of fertile fronds of this sporophyte (seen here ×0.25) develop hollow chambers called **conceptacles** within which are held egg-producing **oogonia** and sperm-producing **antheridia**. Meiosis of these cells is followed by cell division to produce eight eggs per oogonium and 64 sperm cells per antheridium. The eggs and sperm are then released into the water where fertilization takes place, after which the zygote grows directly into the new diploid individual. Compare this cycle with that of *Ulva* (Fig. 4.3) which has two large multicellular generations in its life cycle.

BOX 4.1 KELP: THE TREES OF THE SEA

The kelp are the largest living marine plants. They can grow up to 60 m in length and with their great structural and physiological specializations may be thought of as the trees of the deep.

Typical of the group are members of the genus *Laminaria* (Fig. 4.9a, b), which are common throughout the temperate world, growing up to 2 m in length. They have many adaptations for life in rough subtidal seas. First, they show a great degree of differentiation, not only at the cellular level but also on a much larger scale. Photosynthesis is concentrated in the flat, leaf-like, fronds at the top of the organism, which are strengthened internally by longitudinal cells. These are held up to the light by the buoyancy provided by gas-filled vesicles. The cylindrical stipe below, meanwhile, is well adapted to its mechanical and transport functions, connecting the fronds to the holdfast. It can be

stretched by more than 40% of its length before it breaks. The holdfast is firmly cemented to the floor and its branched form helps prevent it being peeled off the rocks. A further degree of sophistication in *Laminaria* is provided by the positioning of its main meristem at the junction of the fronds and stipe, which allows elongation growth of both. This arrangement has the advantage over an apical meristem that growth is not stopped if the tip is broken off by wave action or eaten by a herbivore. In this respect kelps show convergence with grasses (see Chapter 9) which have a basal meristem that prevents them being killed by grazing herbivores.

Even more spectacular kelps are found in subtidal regions around the Pacific coast of the United States. The giant kelp *Macrocystis* (Fig. 4.9c) and the bull kelp *Nereocystis* (Fig. 4.9d) can both reach lengths of

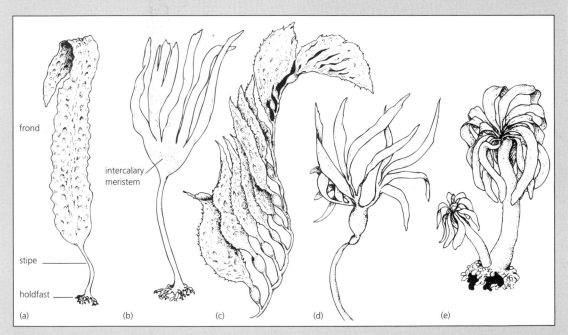

Fig. 4.9 The kelps are the marine equivalent of trees. *Laminaria saccharina* (a) and *L. digitata* (b) are medium-sized organisms found subtidally around the coast of Britain. Note the differentiation into fronds, stipe and holdfast, and the undulating frond in (a) which shows that the plant has grown in sheltered waters. In the huge subtidal forms *Macrocystis integrifolia* (c) and the bull kelp *Nereocystis* (d) from the Pacific coast of North America only the fronds are shown as the stipe can exceed 50 m in length. Note the air-filled bladders used to hold the fronds up to the surface. The sea palm *Postelsia palmaeformis* (e) from California is an intertidal seaweed whose thick trunk can support it even when the tide is out. (All diagrams ×0.07.)

(*Continued on p. 64.*)

up to 60 m. Rather than making them more vulnerable, their great length actually protects them from damage. Their fronds can travel backwards and forwards up to 120 m with waves that would buffet short plants, without their stipes being stretched at all. These seaweeds also show ingenious adaptations to the conditions in which they grow. Plants in sheltered waters have wide, undulating fronds. These not only intercept light well but also increase the turbulence of the water that flows past them, so improving their supply of carbon dioxide and nutrients. Both factors increase the photosynthesis of the fronds and hence the growth rate of the plant. In contrast plants of exposed waters develop flatter, narrower blades, which fold up together in fast flows, reducing drag and helping them resist being torn apart. As we shall see, the growth responses of these giant kelps to wave forces are not dissimilar to the adaptations of trees to the wind.

However, perhaps the most tree-like of all kelps is the sea palm *Postelsia* (Fig. 4.9e) which grows in forest-like clumps in the intertidal zones of the Pacific Coast of California. With its thick, hollow stipe it is capable of standing upright, holding its palm-like fronds up to the light for photosynthesis, even at low tide when it is not supported by water. This kelp even grows like a tree, since its stipe is progressively thickened as it grows. A layer of cells just beneath its epidermis, the **meristoderm**, divides rapidly to produce rings of new tissue, just like the growth rings of trees. *Postelsia* also shows similar adaptations to its environment as do trees. Plants that grow together in clumps shelter and help support each other, like trees in a forest. Consequently each plant can reduce its investment into its stipe and grows taller and thinner in an attempt to outcompete its neighbours for the light.

Of course there are also big differences between sea palms and trees; in particular the gelatinous stipe material is much more flexible than wood. This means that the stipe of *Postelsia* has to be much thicker than that of a tree of the same height. However, the greater flexibility of the stipe also ensures that it can survive waves that would destroy a woody tree. When hit by a wave it can simply bend right over without sustaining any damage, before springing upright again. A rigid tree in the same position would just snap.

All three types of life history are found in benthic algae, although the last is the most common. Part of the variability has resulted because multicellularity has evolved independently several times. Much, though, is probably related to the congenial habitat; in water, spores and gametes can move easily between organisms, and therefore physical factors do not constrain the method of reproduction as much as they do on dry land. The type of life history of a particular alga is therefore affected as much by its ancestry as by its ecology.

Equal development of the haploid gametophyte and diploid sporophyte, known as **isomorphic alternation of generations**, is probably the most common state, and is seen in algae as diverse as the green alga *Ulva* (see Fig. 4.3d), the red alga *Polysiphonia* and the brown alga *Ectocarpus*. However, many algae have independently developed life cycles in which one generation has been greatly reduced. In stoneworts such as *Coleochaete*, only the gametophyte is multicellular; the zygote undergoes meiosis immediately after it is formed. At the other extreme, many red and brown algae have reduced haploid gametophytes. In the brown alga *Laminaria* the gametophytes are reduced to tiny filamentous organisms from whose tip the sporophyte develops directly. *Fucus* (Fig. 4.8) takes this trend even further and, like animals, does not produce a gametophyte at all. Instead it produces eggs and sperm, like an animal, the eggs being held in protective **conceptacles** until they are fertilized; only then is the new diploid individual released to colonize new areas.

Life cycles in which one stage or the other is reduced may have evolved in response to selection pressure to improve the efficiency of reproduction. *Coleochaete*, *Laminaria* and *Fucus* only need to colonize new ground once rather than twice during their life cycle. Since colonization is always a hazardous process, their reproductive losses are thereby reduced.

4.5 ECOLOGY AND DISTRIBUTION OF ALGAE

4.5.1 Advantages and disadvantages of different growth forms

Our brief review of how selective forces have shaped the evolution of the algae seemed to suggest that the larger, more complex algae will have a competitive advantage over smaller, simpler species because they can shade them out. However, they suffer from the disadvantage that, having a relatively smaller proportion of photosynthetic cells, they will grow more slowly and will need more light to survive. Therefore no one type of alga is better in all respects than any other. This example illustrates a fundamental law of biology: adaptations that optimize performance in one set of conditions inevitably reduce performance in others. It is this law which is to a large extent responsible for the great diversity of life and is certainly responsible for the pattern of distribution we see in the algae.

4.5.2 Distribution of benthic algae on rocky shores

Rocky shores in temperate regions typically show an ordered pattern of vegetation (Fig. 4.10), known as zonation, which depends largely on the range of the tide; shores with wider tidal ranges, such as the Atlantic coast of Europe and the coasts of North America, show more widely spread and clearer separation of the species. Large brown algae such as *Fucus* and *Laminaria* dominate the wave-beaten intertidal and subtidal regions, where their greater height

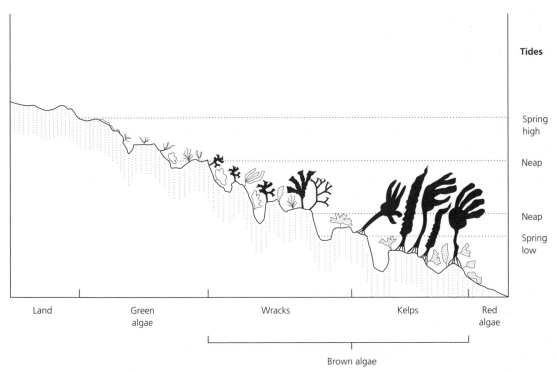

Fig. 4.10 Stylized cross-section through a temperate rocky shore, showing the zonation in seaweeds. Small green algae (clear) dominate rock pools in the upper shore. Below neap high tide the shore is dominated by brown algae (black): the wracks (such as *Fucus* which are found in the intertidal zone, and the larger kelp such as *Laminaria* which dominate the subtidal zone). Red algae (stippled) are found as an understorey beneath the brown algae and as small encrusting forms (too small to show) below the spring low tide mark.

allows them to outcompete the simpler algae. Here they form large 'forests' of wracks or kelps. However, in other areas, the simpler forms have a competitive advantage. Near the high tide mark, where strong wave action and desiccation are major selective forces, only encrusting forms of lichens (see Chapter 5) can survive. In contrast, in sheltered rock pools simple plate-like green algae such as *Ulva* dominate because of their greater growth rate. Simple encrusting red algae dominate deeper seas because they are the only forms which can tolerate the low light levels.

Simpler types of vegetation can also survive within the kelp and wrack forests, just as small herbs can survive in forests (see Chapters 8 and 9). Small red algae like *Laurencia* and *Dumontia* can grow in the low light levels beneath the brown seaweeds, producing an 'understorey' just like that in forests. Similarly, there are several species of epiphytic red algae, including the common form *Polysiphonia* which grows on the fronds of the brown alga *Ascophyllum*. These algae are reminiscent of the epiphytic ferns and orchids which grow on the branches of rainforest trees. There is even parasitism; the fronds of *Polysiphonia* are exploited by the non-photosynthetic red alga *Choreocolax*, just as lianas are exploited by the rainforest parasite *Rafflesia* (see Chapter 8).

Succession on rocky shores

Even in the subtidal areas where more complex forms of algae have the competitive edge, simpler forms may still prosper for a short time if a system is disturbed. If a seashore dominated by wracks like *Fucus* is stripped of its seaweed community by a violent storm, it will first be colonized by simple, faster-growing algae, i.e. diatoms, which form an encrusting mat, and simple green algae such as *Chaetomorpha* and the hollow cylindrical green alga *Enteromorpha*. Only after a few years will these algae be gradually replaced by the climax vegetation, i.e. the slower-growing but more competitive parenchymatous algae such as *Fucus* species, with their understorey of *Laurencia* and *Dumontia*. This pattern of **secondary succession** is strikingly similar, as we shall see, to that which occurs after an area of forest is cut down or burnt (see Chapter 8).

4.5.3 Freshwater algae

Lakes and rivers are in general much calmer and shallower than the seashore. It is therefore no surprise that fresh water, especially lakes and ponds, can so easily become dominated by fast-growing filamentous green algae such as *Cladophora* and *Spirogyra*, which form a characteristic green slime.

4.6 THE CORALS

The low levels of nutrients found in many tropical seas greatly reduces the growth potential of multicellular algae. Their role as the primary producers in these regions is taken over by corals.

4.6.1 Coral structure

Corals are not single organisms, but a symbiotic relationship between two organisms, the coral animals, which are tiny colonial **coelenterates** not unlike Hydra, and spherical golden cells called **zooxanthellae,** which are distributed within their body, and are captive dinoflagellates. The success of the coral symbiosis within tropical seas results from the division of labour between the two symbiotic and mutualistic partners. The coral polyps catch and digest small planktonic animals and also detritus, providing some energy, but more importantly the nitrogen and other nutrients which all organisms need to grow. This is exported to the zooxanthellae which intercept the phosphate that the coral would normally excrete, together with a wide range of nitrates.

The zooxanthellae, meanwhile, which make up around three-quarters of the organic biomass, are the photosynthetic partner. They provide the symbiosis with over 90% of its energy in the form of glycerol which it exports to the coral. By using up carbon dioxide in photosynthesis the zooxanthellae also remove carbon dioxide from the coral and so help it lay down its calcium carbonate skeleton. It is the ability of the coral animals to obtain nutrients from other organisms that allows corals to live in seas in which the nutrient content of the water is so low; it is the ability of the zooxanthellae to photosynthesize that allows them to grow so fast. By combining heterotrophic and autotrophic nutrition, therefore, the coral is able to exploit a nutrient-poor niche, just like insectivorous aquatic plants (see Chapter 11).

4.6.2 Form and ecology of corals

The light requirement of corals both restricts their range and influences their shape. Corals can only grow in the top 60 m of the sea and have to produce structures that can intercept the maximum amount of light. For this reason the shapes and branching patterns of corals are often very similar to those of trees. However, because the resilient calcium carbonate skeletons that are laid down by the coral animals do not rot away, the coral 'forests' are continually growing, like the *Sphagnum* bogs we will meet in Chapter 5. This has resulted in the formation of the great barrier reefs of the tropics, and the coral atolls of the South Pacific, which are all that remain of islands that have long since sunk beneath the waves.

Unfortunately these extensive habitats, which are the basis for the diverse coral reef communities loved by divers and film makers, are now under threat. Some are being destroyed physically by dynamite fishing but much of the problem is due to eutrophication of tropical waters by sewage and agricultural run-off. In the increasingly nutrient-rich waters the corals lose their competitive advantage over the red and brown algae and are increasingly being outcompeted by them.

4.7 POINTS FOR DISCUSSION

1 If land plants died out, which group of the algae do you think would be first to recolonize land and why?
2 Why do you think algae have not evolved seeds?
3 Why don't more seaweeds have stipes that are strong enough to support their own weight, like that of *Postelsia*?
4 We have seen emphasized similarities between seaweeds and land plants. What are the differences?

PRACTICAL INVESTIGATIONS: INVESTIGATING THE DESIGN OF SEAWEEDS

It is possible to learn much about the structural design of seaweeds simply by examining hand sections that you have cut yourself using a single-sided razor blade.

Ideal species to choose as examples of different forms of construction are:
1 the green alga, *Ulva*, the 'sea lettuce' which is a good example of a plate-like alga;
2 the red alga, *Mastocarpus stellata*, which is a good example of a pseudoparenchymatous alga;
3 the brown alga *Laminaria digitata*, which is a good example of a parenchymatous alga. If this is unavailable *Fucus serratus* may be used.

In *Ulva* it may not be possible to cut sections because the plate is so thin; it may be best to examine the surfaces. In *Mastocarpus* it will only be possible to take transverse sections, while in *Laminaria* it is possible to take both transverse and longitudinal sections of the stipe. If the sections are examined under 400× magnification, it should be possible to see the different methods of construction and to identify the different cell types: the small outer layer of photosynthetic cells which contain chloroplasts; the longitudinally elongated internal support cells which are colourless; and, in *Laminaria*, the transport cells which have wider lumens and thinner cell walls and tend to snake down the centre of the stipe.

FURTHER READING

Chapman, A.R.O. (1979) *Biology of Seaweeds*. Edward Arnold, London.
Darley, W.M. (1982) *Algal Biology: a Physiological Approach*. Blackwell Scientific Publications, Oxford.
Denny, M.W. (1988) *Biology and the Mechanics of the Wave-swept Environment*. Princeton University Press, Princeton, New Jersey.
Holbrook, N.M., Denny, M.W. & Koehl, M.A.R. (1991) Intertidal 'trees': consequences of aggregation on the mechanical and photosynthetic properties of sea-palms

Postelsia palmaeformis Ruprecht. *Journal of Experimental Marine Biology and Ecology* **146**, 39–67.

Koehl, M.A.R. (1986) Seaweeds in moving water: form and mechanical function. In: *On the Economy of Plant Form and Function* (ed. T.J. Givnish), pp. 603–634. Cambridge University Press, Cambridge.

Lee, R.E. (1999) *Phycology*, 3rd edn. Cambridge University Press, Cambridge.

Northcraft, R.D. (1948) Marine algal colonization on the Monterey Peninsula, California. *American Journal of Botany* **35**, 396–404.

Raven, J. (1986) Evolution of plant life forms. In: *On the Economy of Plant Form and Function* (ed. T.J. Givnish), pp. 421–492. Cambridge University Press, Cambridge.

Seagel, R.F., Bandoni, R.J. & Maze, J.R. *et al.* (1982) *Plants: an Evolutionary Survey*. Wadsworth Publishing, Belmont, Califormia.

CHAPTER 5

Life on the ground

5.1 INTRODUCTION

The plants we described in the last two chapters successfully colonized most of the niches that are available in the oceans, but what of the dry land? Until about 500–450 million years ago the terrestrial environment that forms the most familiar third of the surface of our planet bore no macroscopic vegetation. This chapter will examine the problems that faced the first plants to venture into this barren scene, and describe the ways in which the first organisms to colonize the land managed to overcome these problems.

We will then go on to consider their present-day descendants, the **bryophytes**, and examine some of the ways in which they have perfected their adaptations to life on the ground. So successful has this group been, that there are more living species of bryophyte than any other land plant group except the flowering plants. They inhabit environments ranging from the hottest deserts to the frozen wastes of Antarctica, festoon the boughs of tropical rainforest trees and form the dominant vegetation in wetlands all over the world. Indeed, as we shall see, they have in many cases been able to compete successfully with the **vascular plants** we will examine in the next two chapters, for life *above* the ground.

The chapter will end with an examination of the parallel world of the lichens: symbiotic associations between fungi and algae which exhibit strikingly similar adaptations for life on the ground to those of the bryophytes.

5.2 LIVING ON DRY LAND

5.2.1 Problems

The oxygen-releasing organisms in the water eventually raised the level of oxygen in the atmosphere to the level of around 1%. This provided an effective shield from harmful ultraviolet light via the atmospheric ozone layer, so the only thing stopping life invading the land was the dryness of the environment. As we have seen, planktonic algae and seaweeds rely on the water around them to keep them moist, to bring them nutrients, to support them, to provide the nuptial swimming pool for their gametes, and to distribute their offspring. On land, the only moisture is held in a thin surface film, which can dry up in times of drought. Land plants are therefore beset by problems not only of survival but also of reproduction and dispersal. Their survival is threatened by the difficulty in obtaining and retaining enough moisture. Even if an ocean-inhabiting plant could survive, its gametes and spores would only be able to swim a few centimetres from their parent along the surface film, so colonization of land would be painfully slow.

5.2.2 Adaptations for survival

The first land plants developed simple but effective solutions to the problems of surviving and reproducing on land. It seems certain that the earliest land colonizers were not modified versions of the most complex of seaweeds, but flat plate-like **gametophytes** of algae, similar to the modern day *Coleochaete* (see Fig. 4.1b, p. 53), with little or no internal differentiation. They overcame problems of

BRICKS RETAIN MOISTURE DUE TO THE AMOUNT OF PORES

obtaining and retaining water by lying flat on the wet surface, and taking up water and nutrients from below. Such plants could rely on simple diffusion for adequate water and nutrient transport and gas exchange.

But the key adaptation that reduced the dependence of plants on a permanent water supply was their development of the ability to withstand desiccation. Like their descendants, the bryophytes, they were probably **poikilohydric**, dehydrating and entering a state of suspended animation during times of drought, and became active and fully hydrated only when their environment was damp (see Plate 1, facing p. 110).

5.2.3 Adaptations for reproduction

The solutions of early land plants to the problem of reproduction and dispersal were similarly amphibious. Their gametes swam to each other just like those of aquatic algae. However, they had desiccation-resistant spores covered with a coat of the waterproof and rigid biopolymer, **sporopollenin**. Such spores, found in rocks over 470 million years old, are the earliest evidence of plant life on land, and may have been dispersed, as are their modern equivalents, by animals or the wind. Equally important in aiding dispersal must have been the development of an upright diploid **sporophyte** stage in the life cycle. This would produce large numbers of spores and would improve their chances of dispersal by holding them up away from the water surface.

5.3 ADAPTATION AND DIVERSITY OF THE BRYOPHYTES

Because they are partly still dependent on a surface film of water, one group of descendants of the earliest land plants, the bryophytes, are not so much fully terrestrial as amphibious. They are most often found in damp environments and are often compared with animal amphibians. But, as we shall see, just like some frogs and toads, some bryophytes can thrive even in the driest conditions such as the most inhospitable deserts.

There are three main groups of bryophytes, the liverworts, the hornworts and the mosses, which diverged over 400 million years ago. All three groups developed a range of adaptations, both in growth and form, and in reproduction, which improved their competitive ability and reduced their reliance on surface water.

5.4 GROWTH AND FORM OF BRYOPHYTES

Many of the liverworts and the hornworts still resemble the plate-like green algae which are thought to be their ancestors, and consist of a flat sheet of cells anchored to the ground by rhizoids. However, competition amongst plants generated more sophisticated structures of two distinct types. The first are the thalli of some liverworts (Figs 5.1 and 5.2), which are organs more than one layer of cells thick, variously refined but nevertheless still plate-like. In the second sort of structures, those of the leafy liverworts and mosses (Figs 5.3–5.5), the plants are differentiated into regular lateral organs attached to a central axis. This arrangement is analogous to that in vascular plants, the shoots of which are differentiated into photosynthetic leaves and supporting stems. All these groups are so distinct they are not thought to have come from the same ancestral line as each other, so their leaves are another good example of convergence.

5.4.1 The thalloid liverworts

As we saw earlier, getting bigger or thicker creates gas-exchange problems and means getting further away from water and minerals. It therefore becomes increasingly difficult for diffusion to maintain the necessary supplies. Some of the most successful thalloid liverworts have highly differentiated forms that overcome these problems. The thallus of *Marchantia* (Fig. 5.1), one of the most widespread genera, is differentiated into two regions with different functions. The lower region is colourless (Fig. 5.1b, c), and bears rhizoids and scales concerned with uptake and transport of water. Some rhizoids run directly downwards into the substrate and convey water and minerals into the thallus. Other rhizoids run parallel to the undersurface and effect

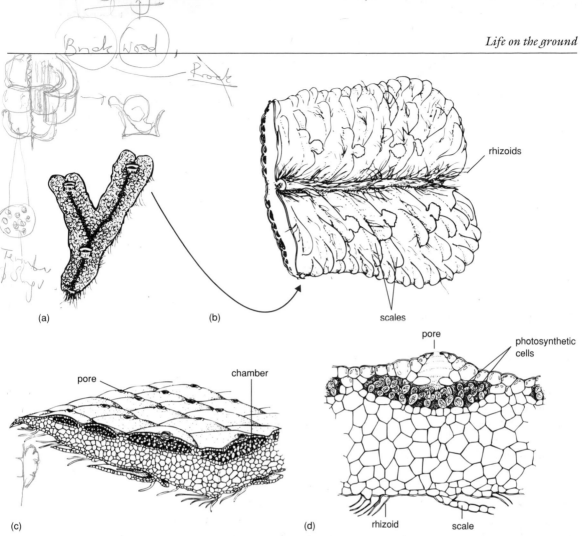

Fig. 5.1 The thalloid liverwort *Marchantia polymorpha*. (a) Surface view (×1.0). (b) Magnified view of the underside. (c) Portion of thallus cut through. (d) Transverse section through the thallus. The lower surface bears hair-like **rhizoids** and flap-like scales, which absorb water and encourage capillary action transport of water around the thallus. The central portion is occupied by nongreen packing tissue, while the upper surface is differentiated into elaborate photosynthetic chambers. The chambers contain files of rounded cells with abundant chloroplasts surrounded by space, which allows efficient gas exchange. The chambers are topped by dome-like rings of cells surrounding a central pore.

transport along the structure and right up to the tips, largely by capillary action. The upper part of the thallus (Fig. 5.1c, d) is green and beautifully designed for photosynthesis. The top is protected from desiccation by a waterproof **cuticle**, and it is perforated by pores which allow gases to diffuse into the thallus. Below lie chambers containing files of cells which are packed with chloroplasts.

The resulting thallus (Fig. 5.1d) looks and works remarkably like the leaves of vascular plants; the structure allows gases to diffuse to the photosyn-

thetic cells while minimizing water loss. However, in *Marchantia* this arrangement is not so efficient as in leaves because, unlike stomata (see Fig. 5.11), the pores cannot be closed. This restricts thalloid liverworts to habitats no drier than the gaps between paving stones, where *Marchantia* can often be found.

5.4.2 Leafy liverworts

Leafy liverworts are usually more delicate than the thick, often liver-shaped, thalloid forms, and are

71

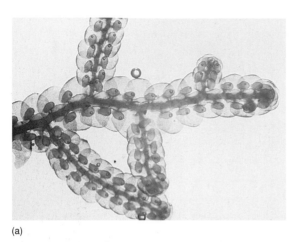

(a)

(b)

Fig. 5.2 Light micrographs showing the morphology of the leafy liverwort *Frullania tamarisii*. (a) General view (×1.5), showing the 'leaves' coming off a central axis. A close-up (b) (×60) shows the characteristic dissimilarity between the leaves. Here one row is modified to form small homes for insects whose excreta and decaying bodies provide the liverwort with extra nutrients.

Fig. 5.3 Scanning electron micrograph of a moss (×15) showing the spirally arranged 'leaves' around the central axis. The channel formed along the centre of the leaf can be seen. It encourages water transport up and down the structure via capillary action.

restricted to damper environments. They are characteristically prostrate, lying flat on their substrates which include the boughs of rainforest trees and woodland soils, and they are subdivided into distinct leaf-like and stem-like portions. These parts, like those of all of the simple plants described in this chapter, are generated from cells carved off from a single pyramidal cell at their apex. In the mosses, which we will consider next, the three rows of leaf-like appendages are often identical in form; in leafy liverworts, the appendages are usually dissimilar.

In the most highly differentiated forms of leafy liverwort, one of the rows becomes highly modified, to serve purposes other than photosynthesis. In some, the organs form water pitchers which help the plants hold onto the water they get (Fig. 5.2). This is important for such delicate organisms because the leaves and stems contain no vascular tissue, and therefore they are inefficient at conducting water internally. Most of the water transport around the gametophyte, as in all bryophytes, is up the outside of the plants. The rhizoids and scales of *Marchantia*

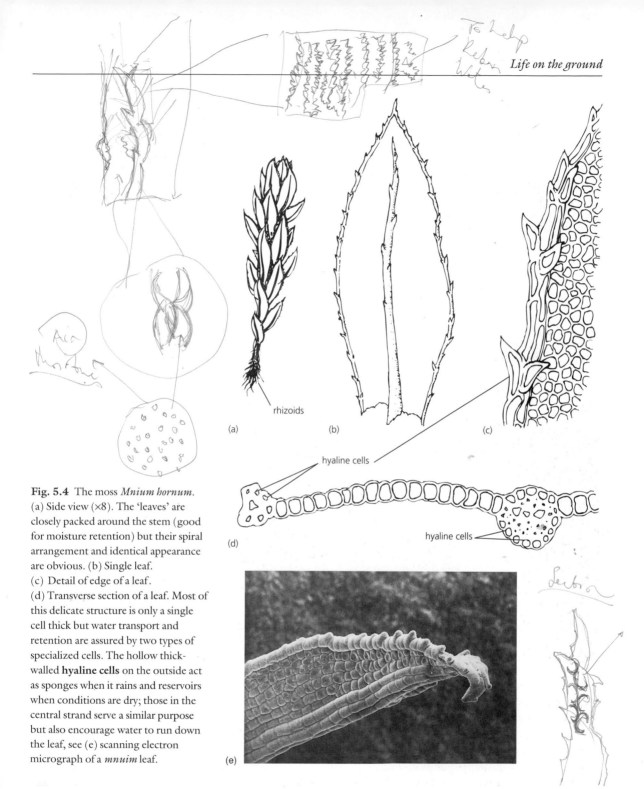

rhizoids

(a) (b) (c)

hyaline cells

(d)

hyaline cells

(e)

Fig. 5.4 The moss *Mnium hornum*.
(a) Side view (×8). The 'leaves' are closely packed around the stem (good for moisture retention) but their spiral arrangement and identical appearance are obvious. (b) Single leaf.
(c) Detail of edge of a leaf.
(d) Transverse section of a leaf. Most of this delicate structure is only a single cell thick but water transport and retention are assured by two types of specialized cells. The hollow thick-walled **hyaline cells** on the outside act as sponges when it rains and reservoirs when conditions are dry; those in the central strand serve a similar purpose but also encourage water to run down the leaf, see (e) scanning electron micrograph of a *mnuim* leaf.

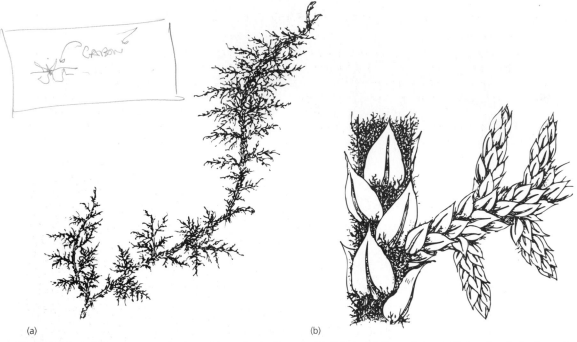

Fig. 5.5 Adaptations to water transport and retention in the moss *Thuidium*. (a) General view (×2) showing the horizontal stem. (b) The feathery branches with their spirally arranged leaves and hair-like stem elaborations which soak up and transport water along the plant.

and the overlapping leaves of the leafy liverworts both help to transport water around these **ectohydric** plants.

Perhaps the most sophisticated leafy liverworts are those in which the third leaf-like organ serves as a home for tiny insects. The tropical *Frullania* (Fig. 5.2) derives benefits additional to those of spore dispersal from its animal associates. The leafy houses protect the insects, and the insects contribute valuable nutrients to the plant via their excreta and decaying bodies. This plant clearly profits from the absence of a thick protective **cuticle**, since the nutrients can diffuse straight into the cells lining the houses.

5.4.3 Mosses

Mosses are the most numerous and familiar members of the bryophytes, with forms ranging from the tiny tough fuzz you might find growing on exposed walls and stones, through lanky tropical forms, to the highly successful but scruffy 'immortal' bog mosses of the genus *Sphagnum* that dominate over 1% of the earth's terrestrial surface (see Box 5.1, Fig. 5.6).

Mosses look superficially like the leafy liverworts, but their identical appendages are arranged spirally around the stem (Fig. 5.3). Few have specialized cells for internal conduction of water; instead most have structures and forms that optimize external transport and retention of water. Although the leaves might appear to have veins or midribs (Fig. 5.4) these lines of specialized cells are strictly superficial modifications which create runways down which water can travel.

Many mosses that have this type of leaf are also able to close them up against the stem in dry conditions, to minimize water loss (see Fig. 5.7 and Plate 1, facing p. 110). This is especially crucial for mosses that grow vertically, some of which have further aids to water retention. Like the xylem of vascular plants, these are often specially thickened cells which are empty and dead when mature. In the bog moss

BOX 5.1 *SPHAGNUM*: THE BASIS OF LIFE IN A BOG

Sphagnum includes plants that form the main component of bogs all over the world. The features that have led to this success include the following.

• A phenomenal ability to take up and hold onto water (Fig. 5.6). The plants can hold 20 times their own weight in water, and water can be drawn up a column of *Sphagnum* plants, helping them form charac-

teristic hummocks and **raised bogs**. This ability also accounts for their use in hanging baskets and, historically, as substitute nappies and wound dressings.

• 'Antiseptic properties'. Wounds of soldiers dressed with *Sphagnum* seldom became infected, thanks to the antimicrobial compounds that reduce fungal and bacterial attack in the damp places inhabited by the plant.

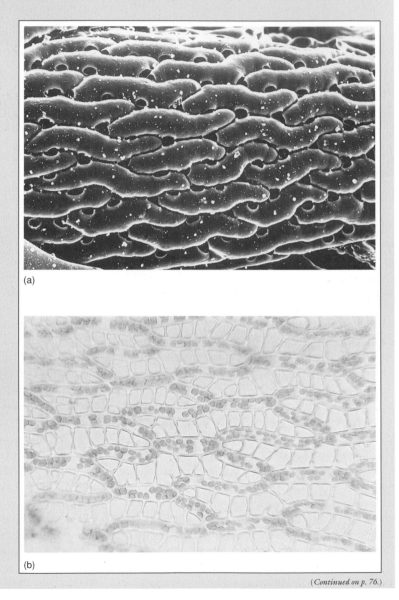

(a)

(b)

Fig. 5.6 Adaptations to water retention in *Sphagnum*. (a) Scanning electron micrograph showing the holes in the dead cells of a leaf, which allow it to absorb and hold water. (b) Light micrograph of a leaf surface, showing the photosynthetic living cells between the colourless dead ones. (Both micrographs ×600.)

(*Continued on p. 76.*)

Box 5.1 contd.

• Efficient nutrient scavenging. Hyaluronic acids in the cell walls of the leaves act like domestic water softeners. Minerals are taken up from solution and hydrogen ions are put in. This ion-exchange mechanism means the plants can obtain all the minerals they need from rainfall, but also renders the water increasingly acidic. This has the added advantage that it creates conditions which are ideal for *Sphagnum*, but less comfortable for most plant competitors.

• An efficient growth strategy. The mature gametophyte lacks rhizoids and is simply a vertical arrangement of stem, leaves and branches to which new parts are added at the top as fast as the bottom-most parts die. The build-up of 'peat' which occurs in some conditions reflects the acidic and often anaerobic conditions around the basal portions, where **saprophytic** microbes, which would break them down, are unable to survive.

• Efficient regeneration. Fragments of plants and even decades-old growth quickly generate new growth in suitable conditions.

'Suitable conditions' is an important proviso in the above sentence, however, because the ascent of man has been something of a disaster for this moss. *Sphagnum* peat is highly prized, both for burning as fuel, and, more recently as a basis for garden and horticultural compost. Unless *Sphagnum* is harvested carefully, which can be expensive, removal of the surface layers can change the hydrology of a bog to such an extent that regeneration cannot occur. Many bogs have already been badly damaged, and environmental groups such as Plantlife encourage gardeners to use composts based on sustainable harvesting or alternatives such as coir, which is a waste product consisting of the outermost layers of coconuts. The Industrial Revolution has also taken its toll, because *Sphagnum* plants are constructed such that they mop up water; they also mop up atmospheric pollutants with the water. The close proximity of the living cells to the water- (and therefore pollutant-) holding cells means that they are unusually sensitive mosses. *Sphagnum* in bogs around areas of high industrial activity continues to decline.

Sphagnum, these **hyaline cells** even have spiral thickenings (see Fig. 5.6, Box 5.1), but they do not form continuous internal tubes. Instead, they are usually the outermost cells—either on the surface which comes to face the dry atmosphere in mosses able to 'shut up shop', or at the tips or edges of the leaves.

The overlapping arrangement of the leaves of mosses also encourages capillary water flow; hair-like extensions of the 'stem' enhance this still further in some of the most successful species (Fig. 5.5). Such arrangements allow some species of moss to become quite large, and essentially independent of soil moisture. They capture their water and nutrients from rainfall, a mode of existence described as **ombrotrophic**. This allows growth to continue from a stem whose basal regions are not in contact with the soil, or which may even be dead. This is not without its disadvantages of course. The topmost plants of large feather mosses, for example, are unable to obtain sufficient moisture from the soil, so they die in the absence of rainfall. The potential for direct entry of dissolved material also renders these mosses particularly vulnerable to atmospheric pollution.

The design of the most sophisticated mosses approaches that of the vascular plants we will consider in the next chapter. Mosses such as the bank hair moss *Polytrichum commune* (Fig. 5.7) not only have multilayered leaves, with files of cells packed with chloroplasts reminiscent of those in the chambers of *Marchantia*, but highly differentiated tissues inside their stems. These internal strands include water-conducting tissue cells called **hydroids** and photosynthate-conducting cells called **leptoids**, which are similar to the **xylem** and **phloem** of vascular plants (see Chapter 6). These mosses lack roots, but their rhizoids are one step up on those of the liverworts, being multicellular and often highly branched. Although these modifications go some way towards making these mosses **endohydric**, they are still far more prone to drying out than truly vascular plants. The toughest mosses are those which further improve their drought tolerance by having the ability to enter a state of suspended animation and an exceptionally efficient repair mechanism for cells damaged by dehydration. So good are these mechanisms that some species can be rehydrated from dry, old herbarium sheets.

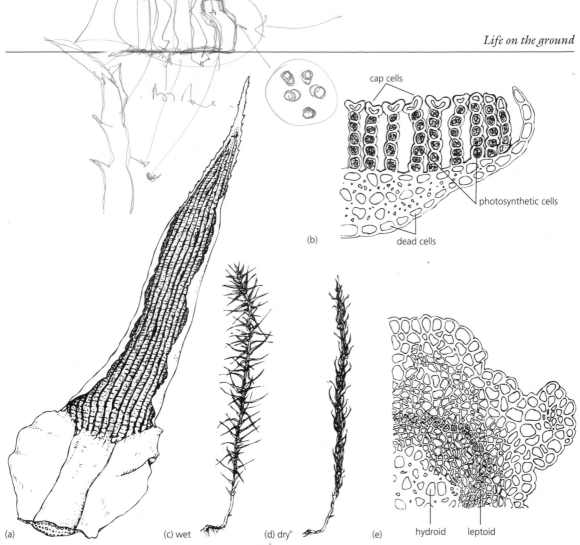

Fig. 5.7 Adaptations to water transport and retention in the bank hair moss *Polytrichum commune*. (a) Single leaf (×20) viewed from above. (b) Transverse section through the leaf (×100). The topmost layer is made up of files of photosynthetic cells topped by thick-walled, hyaline, cap cells which retain moisture. Water is transported along the grooves between the files. In dry weather water is lost from the grooves, causing the leaf to curl up and leaving only the bottom layer of empty dead (hyaline) cells exposed. The thickened central strand also causes the leaf to fold towards the stem when dry. The whole plant therefore curls up from its fully hydrated state (c) to its dry state (d) (both ×1.0), reducing water loss. The lightest of showers will allow the leaves to open out again and re-start photosynthesis. (e) Transverse section of the stem (×60), showing the sophisticated structure. A central strand of large thick-walled cells (hydroids) which are dead at maturity transports water up to the apex internally in these mosses. The smaller thin-walled cells outside this region (leptoids) transport photosynthate down from the topmost leaves to the basal parts of the plant.

5.4.4 Asexual reproduction in bryophytes

We shall see in the next section that sexual reproduction in bryophytes is as reliant on aquatic processes as that of their algal relations. However, bryophytes have a wide range of mechanisms which allow them to colonize new territory without resorting to sexual processes. The phenomenal success of asexual reproduction is evidenced by the findings that large tracts of peat bogs consist of a single genetic individual of *Sphagnum*, and that thriving populations of other mosses are all the same sex—with the opposite sex a continent away!

Simple mechanisms

The mechanisms of asexual reproduction include

77

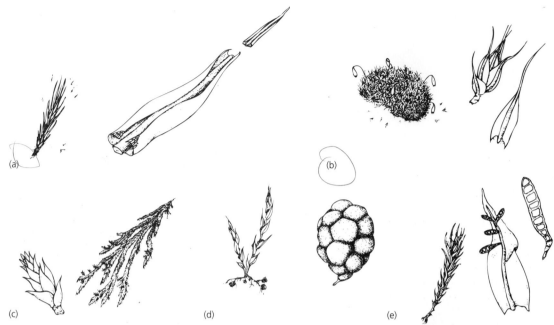

Fig. 5.8 Vegetative reproduction in mosses. (a) *Dicranum tauricum* with leaf tips 'made for' detachment; (b) *Campylopus pyriformis* has fragile shoot tips and entire leaves; (c) *Platygyrium repens* has specialized branchlets; (d) *Bryum rubens* produces gemmae in leaf axils and on rhizoids; (e) *Orthotrichum lyellii* produces them down the centre of the leaves. All these bodies can reproduce (clone) a whole parent plant.

growth and fragmentation. Fragments may be accidentally detached from the parent, the base of the parent may decay to leave several individuals, or new plants may be formed in a very deliberate manner. Some mosses produce deciduous branches or even shoots, which look like a miniature version of a parental shoot and whose role is to detach and propagate the parent. Running a finger gently over colonies of species with this ability produces a forest of tiny shoots well able to adhere to passing animals or blow away in the wind (Fig. 5.8). Regenerative ability is very high in bryophytes, and perhaps the best evidence for this is that even tiny fragments of leaf tips are capable of growing into entire new plants in some genera (Fig. 5.8a).

Gemmae

Many bryophytes produce more specialized propagules called **gemmae**. Like a deciduous shoot, each gemma is the product of mitotic cell division, so is genetically identical to the parent. While attached to

the parent, a growth inhibitor travels up into the gemma through its slender stalk to prevent it growing and smothering the parent (Fig. 5.8e). Gemmae are often produced in special structures such as the gemma cups illustrated in Fig. 5.9. The design of such cups creates a funnel which directs raindrops rapidly down into the base, where the force detaches the stalk connecting the gemma to the parent. Fortunate gemmae get raindrop-blasted into the air and land away from the parent, where, having been freed from inhibition, they start to grow. Other species generate special poles with gemmae on top, which doubtless facilitates the action of animal vectors, such as slugs, which are known to disperse gemmae (Fig. 5.9).

5.4.5 Sexual reproduction in bryophytes

The basic life cycle

All the structures we have considered so far have been gametophytic, yet we have not come to the products which are responsible for this name—the

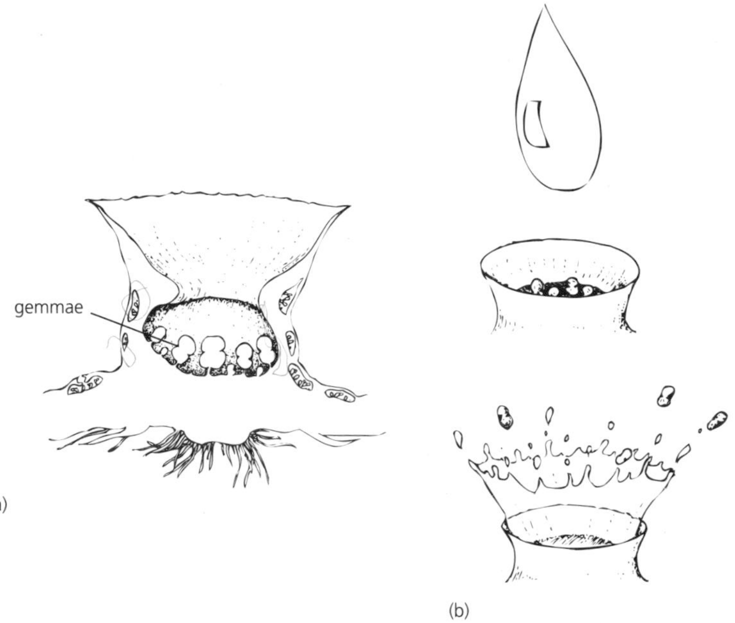

Fig. 5.9 Transverse section through a gemma cup of *Marchantia polymorpha* (×10) showing the gemmae—bilobed miniatures of the parent plant, ready for blast off. Each is attached at the base by a slender stalk. (b) Before and after a raindrop hits a gemma cup of *Marchantia polymorpha*. The force of the drop breaks the slender attachment stalk and the gemmae are violently launched into the air, hopefully touching down some distance from the parent plant.

gametes. This is no oversight, since bryophytes spend the majority of their existence as gametophytes; indeed many living species have never been known to reproduce sexually. When they do reproduce sexually, bryophytes produce multicellular sexual organs (**gametangia)** and the process in the different groups shares many similarities. The male organs, called **antheridia**, produce small biflagellate male gametes which require a film of moisture in which to swim to the mouths of the multicellular structures, called **archegonia**, which house and protect the larger, non-flagellate female gametes (the eggs). After fertilization, the zygote divides mitotically to produce a diploid sporophyte which never becomes independent of the parent gametophyte, and usually relies heavily upon it for protection, support and nutrition (Fig. 5.10).

Sporophytes of liverworts

Most bryophyte sporophytes have simple stalks which bear terminal structures called **capsules**, containing the spores (Figs 5.10–5.12, Plate 3, facing p. 110). However, some of the most notable evolutionary novelties of the three bryophyte groups occur in their sporophytes, and are related to their function

of releasing their spores well away from the surface film of water. Thalloid liverworts possess tiny sporophytes (Fig. 5.12), which, in plants such as *Marchantia*, are held out below umbrella-like extensions of the gametophyte. These extensions eventually raise the capsules from the surface, but not far away from the water film that clings to the gametophyte. The release of the spores away from the film is aided by a second type of cells, the **elaters** (Fig. 5.13), which are also held in the capsule. As the capsules dry out, water is lost from these cell, whose helically thickened walls are compressed like tiny springs. When the capsule is dry it breaks open, the film of water within each elater breaks and they spring open, explosively catapulting the spores clear of the capsule.

Sporophytes of hornworts and mosses

The sporophytes of hornworts and particularly mosses (Figs 5.10 and 5.11) are much larger than those of liverworts. They raise the spores well above the water film, and possess adaptations which allow them to grow without becoming desiccated. They are coated by a water-tight **cuticle**, while the supply of oxygen to internal tissues is aided by special pores

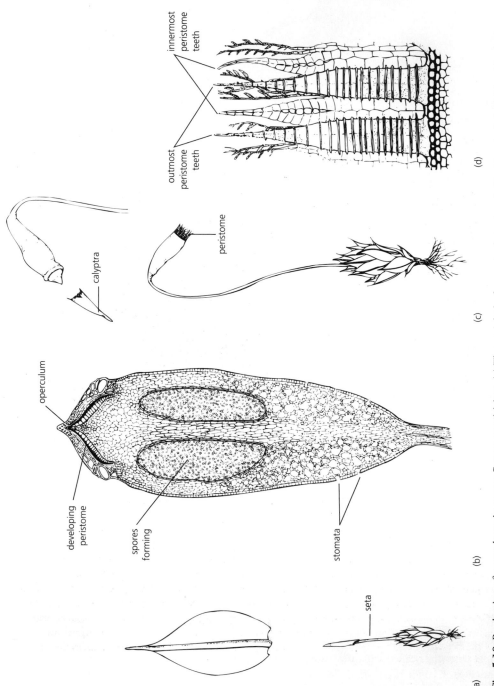

Fig. 5.10 Production of sporophytes in a moss *Bryum caespiticium*. (a) The upright leafy gametophyte stage (×2) with a single leaf alongside. At the very base are the tiny, multicellular rhizoids which anchor the plant, then a stem surrounded by tightly overlapping leaves, at the apex is a developing sporophyte. The stem (seta) has not fully elongated and the capsule is still protected by a sheath, called the calyptra, which was originally part of the archegonium wall (covering the egg cell). (b) Section through the topmost portion of the same stage (×20) with the calyptra removed, showing some of the complex beauty of the sporophyte. The capsule wall has stomata and the spores are formed in a sac which is protected by a multicellular covering. As the sporophyte matures, the calyptra drops off to expose a lid (operculum), which protects the delicate peristome teeth until (c) that too detaches, and changes in humidity cause the two rows of peristome teeth (d) to flex to and fro and release the spores.

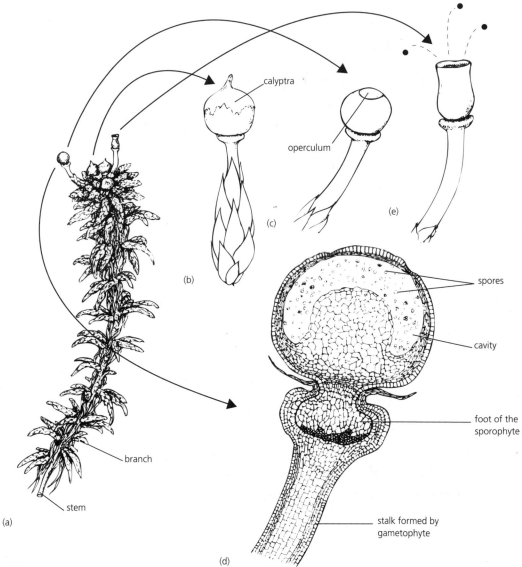

Fig. 5.11 The bog moss *Sphagnum papillosum*. Compare these drawings with those in Fig. 5.10. (a) Topmost part of a gametophyte (×1.5)—the living material at the surface of the bog. Beneath is more stem and branches, which are dead at the very base (but there are no rhizoids in the mature plant). At the apex is a cluster of branches, amongst which are several sporophytes. The youngest still have their calyptras (see b). When the calyptra has fallen off, the rounded capsule can be seen to have a thinner region circling around the top, which produces a lid or operculum (c). Inside the capsule (d) it is clear that the spores lie in a cavity just underneath this lid (and that the stalk is part of the gametophyte, with only the foot-like base as the non-capsule part of the sporophyte generation). As this structure dries out and loses moisture, pressure builds up on the spore sac (imagine squeezing a bottle of washing-up liquid with the cap still on) until eventually it becomes too much for the lid, which blasts off (with an audible pop) and the spores explode out (e).

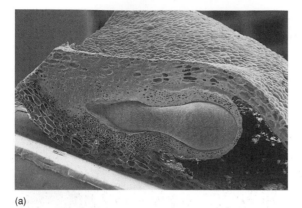

(a)

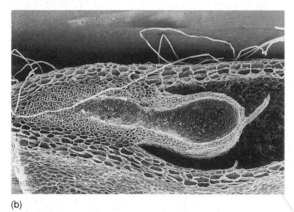

(b)

Fig. 5.12 Scanning electron micrographs (×7) showing tiny developing sporophytes of the liverwort *Pellia*, still within the adult gametophyte. (a) Shows a sporophyte within its protective covering; (b) shows the sporophyte has been removed to show the multicellular protective layer formed by the archegonium (calyptra) which surrounded it.

called **stomata** (Fig. 5.14). Stomata are an elegant solution to the problem of supplying essential gases while avoiding excessive water loss. They have one or more specially modified cells around a pore which allow it to be opened or closed. The plant can control the movement, so when water is plentiful the stomata can be opened wide, whereas when water is scarce, the stomata close and conserve moisture.

The sporophytes of mosses also possess a final adaptation. Their tissue is supplied with water and nutrients from below by a central rod of water-conducting tissue similar to **xylem**. As we shall see in the next chapter the three major adaptations of

some moss sporophytes—their internal conducting tissue, cuticle and stomata, are also keys to the success of the **vascular plants**.

Spore dispersal in mosses

Apart from their height, moss sporophytes have other adaptations to aid spore dispersal. Their capsules lack elaters, but most have ingenious devices to release spores. Most mosses rely on wind dispersal, and have a broad range of capsule structures and spore release mechanisms. *Sphagnum* uses an explosive mechanism (Fig. 5.11), but most other species have more elaborate capsules that release spores more gradually. The structures that effect this are called **peristomes**. Some function like pepper pots—the capsules open and close in response to changes in humidity. This causes the rows of teeth of other species to flex and move (Figs 5.10 and 5.15, Plate 3, facing p. 110). The presence of bryophytes on Antarctic mountains and newly emerged volcanic islands testifies to the efficiency of long-distance dispersal of spores.

But perhaps the most fascinating dispersal mechanisms are those of the dung and carrion mosses. The capsules of these mosses include brightly coloured rings of tissue, which make them look, and function, rather like tiny flowers. They open to reveal a stalk bearing sticky masses of material containing the spores. Dung or carrion flies are attracted by the colours, and odours of the capsules in some species, and serve as animal vectors for the dispersal of these plants to fresher dung or carcasses.

5.5 LICHENS

5.5.1 Evolution and distribution

Having a thalloid growth form is such a successful strategy for surviving on land that it has also been adopted by a totally unrelated group of organisms, the lichens. Lichens are not strictly plants, but intimate associations between a fungus and one or more photosynthetic partners. The chlorophyll-containing partners (**photobionts**) include members of the green algae and the cyanobacteria. The fungus

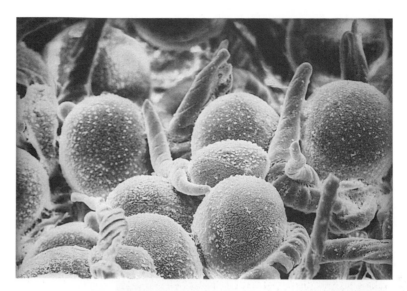

Fig. 5.13 Scanning electron micrograph (×60) of the spherical spores and long helically wound **elaters** of a liverwort. The elaters are shortened as they dry by the surface tension of the water between the spiral thickenings, but the water film eventually breaks and the unwinding of the elater 'springs' fire the spores into the air.

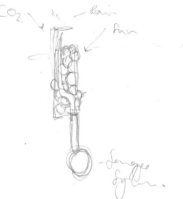

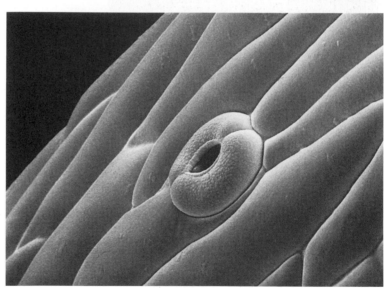

Fig. 5.14 Scanning electron micrograph (×1500) of the epidermis of a plant, showing a **stoma** with the two guard cells which control carbon dioxide uptake and reduce water loss.

obtains not only an organic form of carbon from the partnership, but, if they harbour cyanobacteria, also a ready-fixed form of nitrogen. The photobionts gain a water-retentive covering which permits the survival of species that would otherwise perish in most of the terrestrial conditions in which lichens are found. The fungus may also protect the photobiont from photo-damage during dry periods by becoming opaque.

There is some evidence that lichens were amongst the earliest colonizers of dry land, and they certainly thrive even today on such harsh environments as bare rock. Lichen-like associations seem to have arisen many times during evolutionary history, indicating that the benefits to the organisms involved must greatly outweigh the disadvantages (such as the destruction of some algal cells through penetration by over-enthusiastic fungal hyphae). There is great diversity and lichens inhabit an impressive range of environments (see Chapter 10). They may be found deep within Antarctic rocks. They thrive in the frozen Arctic tundra, where 'reindeer moss' *Cladonia* (see Plate 3, facing p. 110) provides the

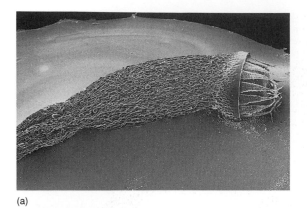

(a)

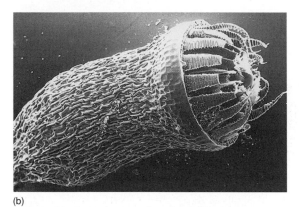

(b)

Fig. 5.15 Scanning electron micrographs (×10) of the capsule of a moss, showing the tooth-like peristomes that control spore release. They open in dry periods when dispersal will be most efficient.

sole fodder for reindeer for much of the year. Some other lichens even cope with the blistering heat of deserts (see Chapter 10). Most of the credit for the survival of lichens in environmental extremes goes to their potential for 'suspended animation'. Like the desiccation-tolerant bryophytes, lichens shut up shop when water becomes limited and survive in a dehydrated state. Rehydration may be delayed by months or even years, but brings with it a rapid return to normal metabolism.

5.5.2 Structure

The species name of a lichen reflects its **mycobiont**, as the form of the association is determined primarily by the fungus, which usually makes up about 90% of the biomass. The remaining 10% is composed of the photobiont/s, which can be found within the lichen. The photobiont often forms a layer of green tissue just beneath the surface (Fig. 5.16), calling to mind the mesophyll of a leaf beneath its protective epidermis or the placement of the photosynthetic cells in *Marchantia* (see Fig. 5.1). The layer on the outermost surface, however, is quite unlike the cuticle of a vascular plant, being composed of fungal hyphae bound together with a proteinaceous matrix. Although this provides a certain amount of physical protection, it is far from water-tight. This is good news for the cells underneath in some ways, since this aids uptake of water during wet weather and hence speeds up the rehydration process. However, it has certain disadvantages. For a start it means that water can be rapidly lost in dry weather. Together with the small volume of photosynthetic tissue, these large swings from wet to dry mean that many lichens grow very slowly indeed. Some of the simplest crust-like forms have even been used to date ancient stones and monuments in the science of **lichenometry**.

A second result of the leakiness of lichens is that just as water is able to flow in through the surfaces and, in some cases, up the root-like **rhizines**, so are dissolved materials. This provides an efficient way of taking up mineral nutrients, but it is a big disadvantage in polluted areas. Acid rain, for example, kills all but the toughest lichens (and bryophytes), and therefore the lichen flora of an area provides an excellent indicator of its air quality. However, lichens are surprisingly tolerant to heavy metals and radioactive pollutants, so they can take up large amounts of radionuclides but still continue to grow. This is useful when tracking the flight paths of space capsules that fall to earth, or prospecting for uranium, but proved tragic for the Lapps of Scandinavia. Fallout from the Chernobyl nuclear reactor accident in the 1980s accumulated in the bodies of their lichen-feeding reindeer, rendering them too dangerous to eat.

5.5.3 Reproduction

The attractive lichen *Cladonia floerkiana* is an ideal form from which to consider the problem of repro-

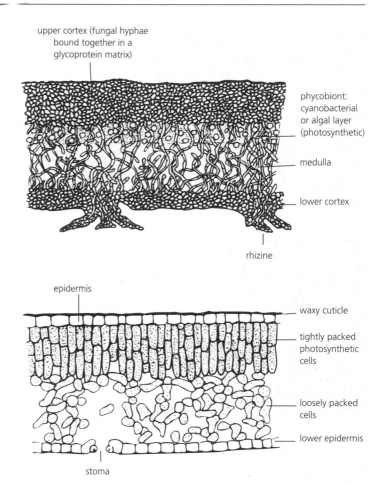

upper cortex (fungal hyphae bound together in a glycoprotein matrix)

phycobiont: cyanobacterial or algal layer (photosynthetic)

medulla

lower cortex

rhizine

epidermis

waxy cuticle

tightly packed photosynthetic cells

loosely packed cells

lower epidermis

stoma

Fig. 5.16 Diagrammatic representation of a section through *Lobaria*, a lichen (above) and the leaf of a vascular plant, *Malus* (below), which show some similarities (both ×50).

duction. Its most obvious feature is its tower-like reproductive organs (see Plate 3, facing p. 110) which are strongly reminiscent of the sporophytes of bryophytes. Of course to reproduce something in which there are two or more organisms ideally requires that they are all included in the process. Strangely, however, the red hat-like organs on the tops of the towers generate and release only fungal spores. The jam-tart like structures on some other lichens have been shown to include some algal cells along with the fungal spores, but in most cases the fungi must rely on (re)finding the right partners.

Lichens do have options other than fungal spore production, however. Simple fragmentation is probably the most common way of reproducing, but many species generate specialized propagules containing both phyco- and mycobionts. Some produce tiny balls called **soredia** (Fig. 5.17) which comprise algal cells wrapped in fungal hyphae, and which may be easily blown to pastures new. Others produce rather larger, finger-like projections called **isidia** which snap off when branches rub together or animals brush past. Once again we have a group of organisms in which some members get around with the help of the wind, while others use the assistance of animal vectors.

5.6 POINTS FOR DISCUSSION

1 Why do so few bryophytes produce internal water-conducting material in their gametophytes?

2 What might prevent bryophyte trees 30 m or more in height from evolving?

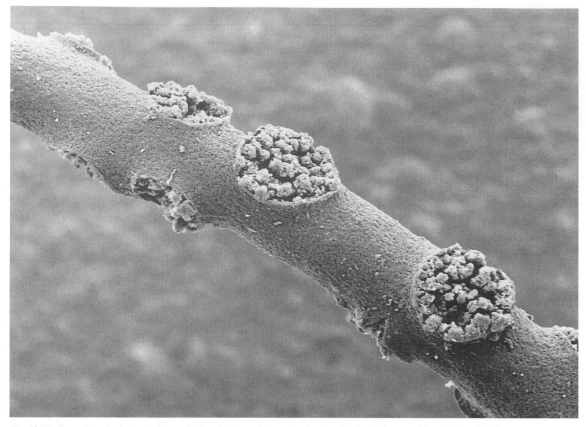

Fig. 5.17 Scanning electron micrograph showing soredia production by the Spanish moss lichen *Usnea* (×15).

3 What are the relative advantages and disadvantages of the thalloid and leafy forms of bryophytes? How might this affect their distribution in the environment?

4 Why are there no lichens with a morphology like leafy liverworts or mosses?

FURTHER READING

Boron, G. (1991) *Understanding Lichens: a Beginner's Guide*, Richmond Publishing Co, Richmond.

Carlile, B. (2000) For peat's sake. *Biological Sciences Review* **12** (3), 19–22.

Daniels, R.E. (1989) Adaptation and variation in bog mosses. *Plants Today* **1**, 139–144.

Ingold, C.T. (1939) *Spore Discharge in Land Plants*. Clarendon Press, Oxford.

Nash, T.H. (1996) *Lichen Biology*. Cambridge University Press, Cambridge.

Paton, J.A. (1999) *The Liverwort Flora of the British Isles*. Harley Books, Colchester.

Richardson, D.H.S. (1981) *The Biology of Mosses*. Blackwell Scientific Publications, Oxford.

Stewart, W.N. & Rothwell, G.W. (1993) *Palaeobotany and the Evolution of Plants*, 2nd edn. Cambridge University Press, Cambridge.

Life above the ground

6.1 INTRODUCTION

Although the bryophytes successfully colonized the land, we saw in the last chapter that most remained partially amphibious. Because their dominant gametophytes rely on a surrounding film of water to keep them moist, they cannot grow very far away from their substrate except in a damp mass. They tend to be low-lying and are vulnerable to being shaded out.

The group of plants that has come to dominate the land, the **vascular plants**, have a very different life history, despite being closely related to the bryophytes. The dominant stage in their life cycle is not a plate-like gametophyte, but a tower-like **sporophyte** which has adaptations similar to those of some moss sporophytes and can stand up without becoming desiccated. The sporophyte has a waterproof **cuticle**, which reduces water loss, **stomata,** which allow rapid gas movement to the internal cells, and internal conducting and support tissue—**xylem**. The walls of xylem cells contain cellulose fibres in a hemicellulose matrix, but they are laid down in a greater number of thicker layers than those of ordinary cells. They are further strengthened by cross-linking of the cellulose within the cells with a material called **lignin**. Interconnected files of xylem cells, which die and become empty at maturity, create conduits through which water and nutrients can pass upwards far more quickly and easily than by diffusion (Fig. 6.1c). Xylem not only improves water transport. Because the cell walls are strengthened, it is also a very rigid material, and helps support the sporophyte, enabling it to grow taller and disperse its propagules more efficiently.

In mosses the sporophyte is a largely parasitic structure; it obtains sugars and most of its nutrients from the gametophyte and its only function is to produce and disperse the spores. The sporophytes of vascular plants are the main stage in the life cycle and they live independently, and so are different in three main respects from moss sporophytes. They contain abundant photosynthetic cells, which enable them to produce all their own food; they have underground organs to absorb water and nutrients, and they possess **phloem** tissue which transports the sugars they produce down to the less illuminated parts of the plant, just as in the seaweeds and moss gametophytes we saw in Chapters 4 and 5.

The major advantage to vascular plants of being desiccation-resistant **endohydric** structures is that they can grow taller, and hence shade-out competing bryophytes. They are also less dependent on a surface film of water, and therefore can more easily inhabit drier habitats. As we shall see, much of the subsequent evolution of vascular plants has resulted from further competition between themselves for light and from adaptations which reduced still further their dependence on moisture.

6.2 THE FIRST VASCULAR PLANTS

Perhaps because they were larger and tougher, we have a far better fossil record of vascular plants than of bryophytes. The earliest vascular plant fossils date from the late Silurian period, about 420 million years ago, but one of the plants for which we have the most beautifully preserved and complete fossil records is *Rhynia* which dates from the Devonian period around 400 million years ago. The plant is named

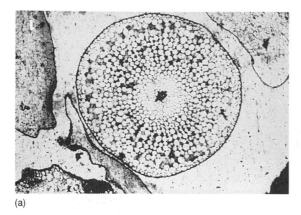

(a)

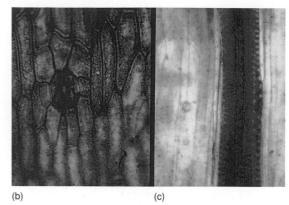

(b)   (c)

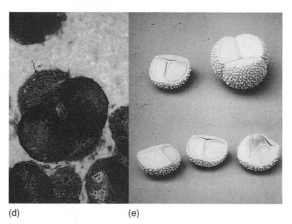

(d)   (e)

Fig. 6.1 Micrographs of *Rhynia* fossils. (a) Transverse section of the stem (×10), obtained by cutting slices through the chert rock in which it is fossilized and then grinding them down to be thin enough to see cellular detail. The tiny strand of xylem in the centre, larger parenchyma cells and the cuticle on the outside are all clearly visible. (b) Epidermis with stoma. (c) Longitudinal section of the vascular strand showing a xylem cell with its spirally wound thickenings (stained dark) and nearby phloem (both ×600). (d, e) The spore tetrads, shown in the original fossil (d) (×500) and in reconstruction (e). The latter shows how the faces of the spores in contact with each other generate a Y-shaped mark on each, and illustrates the thick protective sporopollenin-rich covering over each spore.

after the village of Rhynie in Scotland, near which are chert deposits containing plant remains which are believed to represent a marsh community inundated by hot mineral-rich water from volcanic events. The resultant petrifactions show plants preserved in their original positions with superb detail of all the cells and tissues (Fig. 6.1a).

6.2.1 Structure

The fossils clearly show that *Rhynia* plants had cuticle, perforated by stomata on their surfaces and a strand of xylem up the centre of the stems (Fig. 6.1). Cutting sections through the rock allows the reconstruction of entire plants, and Fig. 6.2 shows what *Rhynia* plants must have looked like. We assume that they were green, because all the ultrastructural, biochemical and molecular evidence from modern-day land plants indicates their affinities with the

Chlorophyta. They had no leaves or roots. Instead, their basal parts were horizontally growing stems anchored, probably in mud, by simple unicellular root-hair like **rhizoids**. Their aerial parts were slim, leafless, upright stems, each supplied with water through the narrow central strand of xylem, but supported largely by the turgor pressure of the outer cells, just as in modern dandelion flower stalks. Some stems bore reproductive structures called **sporangia** (Fig. 6.2).

6.2.2 Spore production

Although we cannot determine exactly how the first land plants reproduced, it is clear that they produced spores coated with desiccation-resistant sporopollenin, just like modern mosses. These were formed in groups of four or **tetrads**, just as in many living plants (Fig. 6.1d, e). The propagules were probably

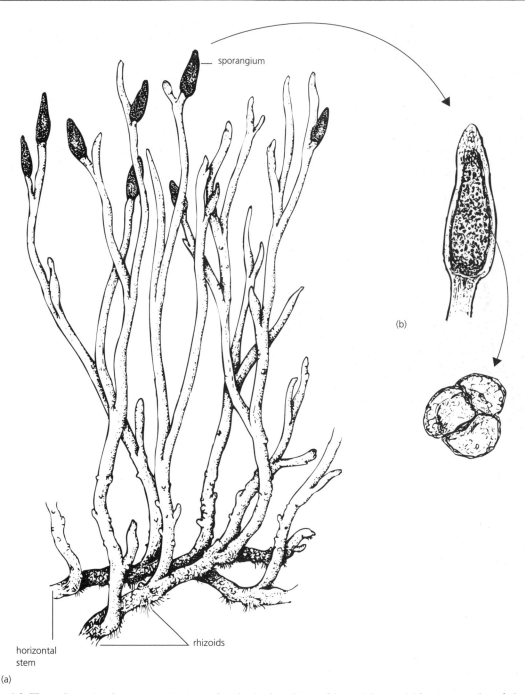

(b)

(a)

Fig. 6.2 Three-dimensional reconstruction (×1) of a *Rhynia* plant. Some of the upright axes (a) bore sporangia at their tips, seen in more detail in (b).

dispersed via the two mechanisms still used by present-day groups of land plants: wind and animals. The position of the spore-containing structures (sporangia) at the topmost region of the stems, and the nibbled appearance of many of them, indicate that both mechanisms may have been employed. It seems likely that the employment of animals as agents of dispersal of propagules was a very early evolutionary step. Most of the spores of present-day plants fed to insects pass through them unscathed, protected by the sporopollenin coat; in natural conditions they would be deposited some distance from the parent plant in a nutrient-rich substrate.

6.2.3 Life cycle

We have no idea of the rest of the life cycle of *Rhynia* but there have been some fossils discovered recently in the Rhynie chert that suggests that such plants probably had gametophytes unlike any plant alive today. The evidence points to gametophytes that may have had vascular tissue, and which were similar in size to their sporophytes—a condition described as **isomorphic** (see section 4.4). However, any gametophyte growing well away from the soil film surface would have problems with reproduction; it would be difficult for the gametes produced by the tips of the axes to swim away from the plant. For this reason there was, probably even early on in plant evolution, strong selection pressure for reduction of the gametophyte. This has certainly been the case during the evolution of the modern plant which bears an uncanny resemblance to *Rhynia*, the 'whisk fern' *Psilotum* (Fig. 6.3). When the spores of *Psilotum* germinate, they produce a gametophyte that looks quite like part of the parent sporophyte (Fig. 6.3d) but is far smaller and grows completely underground in wet regions of the tropics and subtropics. In this wet environment, the tiny flagellate gametes produced by the gametophyte can readily swim to another one and so carry out sexual reproduction. The process is just like that in the bryophytes we saw in the last chapter. The need for water for reproduction is not, however, a serious impediment to *Psilotum*, because it manages to compete very successfully with other plants, even becoming a weed in some greenhouse conditions.

6.2.4 Associations with fungi

Among the many similarities between *Psilotum* and *Rhynia* there is one that has proved a final important component of adaptation to life on land: the presence of a mutualistic fungus. Inside the gametophytes and horizontal portions of the sporophytes of *Psilotum* is a zygomycete fungus. Fungi require carbon in an organic form, so we usually associate fungi with decay or disease, but at least for the gametophyte of *Psilotum*, the association is the key to health and life. Being subterranean, the gametophyte cannot use sunlight to fix carbon dioxide into organic carbon. It turns the tables on its symbiont and obtains carbon from the fungus. Later on, when gametes have met and a sporophyte has formed, organic carbon flow within the association reverses. The upright photosyntheic stems of the sporophyte provide the organic carbon needed for the growth not only of its underground rhizomes but also the fungus. Fossils of the horizontal lowermost portions of *Rhynia* show clear evidence of similar fungi.

Association with fungi was clearly a very early evolutionary step, and almost every vascular plant alive today has an association with one or more fungi which is mutualistic in nature. Mutualism is a form of symbiosis where both of the organisms benefit, and in the case of plants and fungi that form **mycorrhizal** associations the relationship can be a critical determinant of survival. The fungi extend the absorptive surface area of roots and therefore help them to exploit a greater volume of soil. This allows plants to survive in soils poor in nutrients such as nitrogen or phosphorus, or subject to drought. The fungi also provide protection from parasitic fungi that might otherwise invade and kill the plant. The earliest land plants almost certainly encountered nutrient-poor substrates so there would have been strong selective pressure for the development of mycorrhizal associations.

6.3 ADAPTIVE RADIATION OF SPORE-PRODUCING PLANTS

Rhynia was clearly perfectly able to survive in the terrestrial environment, but because each stem was just

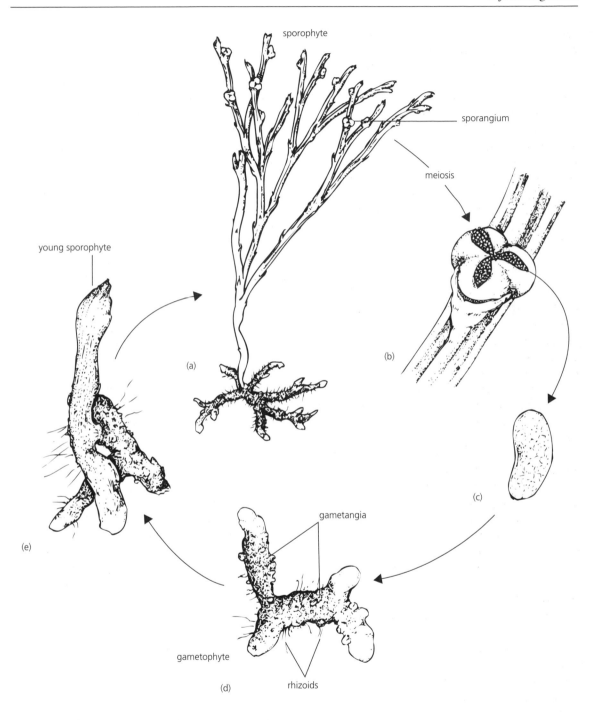

Fig. 6.3 Life cycle of the whisk fern *Psilotum*. (a) The diploid sporophyte, which is the dominant generation of the life cycle as in other vascular plants. Although the plants lack proper leaves, they do have tiny projections off the stems (which might be how microphylls started to evolve in early lycopods). The simple sporangia (b) split open to release spores (c) that germinate and grow into the subterranean gametophyte generation (d). When a motile male gamete successfully fertilizes one of the eggs in the archegonia, an embryo starts to form which eventually pushes a shoot up through the soil and completes the cycle.

91

a simply branched cylinder only 20–30 cm tall, it would have produced little shade, and would itself have been vulnerable to shading. Having no roots, it would also have been a poor competitor for water and nutrients.

The later part of Devonian period certainly saw the demise of the majority of rhyniophytes, which were replaced by major groups of plants that competed more effectively for light, water and nutrients. As we shall see, in fewer than 50 million years these new groups independently developed several major adaptations in their vegetative structures. They mastered the art of reaching up and spreading out, by developing thick, strong stems, to become **arborescent** (tree-like) structures. They improved their capture of light by developing increasingly complex lateral appendages—**leaves**. And they improved their anchorage and absorption by developing true **roots**.

The efficiency of reproduction of vascular plants gradually improved, culminating in the production of different spores (finally becoming exclusively male and female) and smaller gametophytes (which could be protected within the spore coat, or even the host plant).

The rest of this chapter outlines some of the important steps taken in the evolution of growth and form and of reproduction in the competing groups of vascular plants during the Devonian and Carboniferous periods and introduces some of the survivors of the competition.

6.4 GROWTH AND FORM OF VASCULAR PLANTS

6.4.1 The club mosses (Lycophyta)

The lycophytes are a group of plants that proved extremely successful throughout the Devonian and Carboniferous periods, but that are represented today by only a few small genera, which seldom dominate their ecological niches.

Microphylls

The lycophytes are commonly called club mosses, not because they are related to mosses, but because some resemble mosses at first glance; others have club-like reproductive organs. The moss-like look is due to one of the major advances achieved by this group—the simple leaves which surround the stem and enhance light capture. Unlike those of mosses, however, the leaves are supplied by a single strand of vascular tissue and are referred to as **microphylls**. *Micro* means small and *phyll* means leaf, and the present-day lycopods certainly do have small leaves, but microphylls have not always been small. Back in the coal-forming forests of the Carboniferous were enormous arborescent lycopods with very long (up to 40 cm) microphylls (Fig. 6.4a). Although large, these leaves were simple—just a tapered linear structure with a single vein. Microphylls are therefore simple, not necessarily small, leaves which are thought to have evolved from tiny lateral projections of stems. The origins of the more complex leaves in other groups are thought to be quite different, as we shall see in the next section.

Secondary thickening

The second major advance of the group, one that allowed them to grow taller and produce the world's first forests, was their ability to lay down extra thickened material as the plant grew, and so improve their strength and water conduction. This allowed them to raise themselves above their competitors, where their leaves could capture the light unhindered. As in *Rhynia*, the central strand of woody xylem would have provided only a small amount of support for the stem. The strength of the first trees was therefore augmented by the progressive laying down of bark-like cortex around the outside of the stem. Water transport was also improved by a ring of cambial tissue which laid down a limited amount of secondary xylem in rings around the outside of the primary vascular strand. This was an early form of **secondary thickening** analogous to that produced by modern trees (see Chapter 7). These plants were thereby able to develop into huge trees like *Lepidodendron* (Fig. 6.4b) which grew to heights of up to 40 m.

Anchorage structures

Of course a huge stem is no good if it has nothing to anchor and stabilize it, and the development of enor-

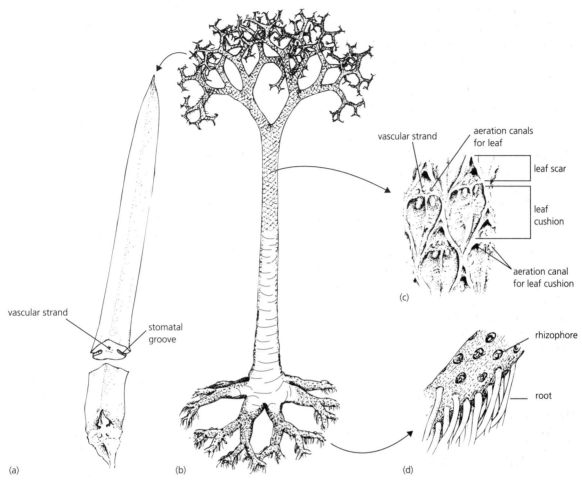

Fig. 6.4 Artist's reconstruction of the arborescent lycopod *Lepidodendron*, pieced together from fossils collected from Carboniferous rocks. (a) Single **microphyll** (×0.25). Although these were 40 cm long, a simple single vein was all that transported water up and photosynthates down. (b) The whole plant (×0.003). The tall trunk and branches were covered in elaborate scars easily visible in close-up (c) left behind when the microphylls fell off. The air channels that ran through the plant can also be seen. (d) Portion of a rhizophore with attached roots that anchored the plant and scars left behind by old roots.

mous rooting systems was the third advance which contributed to the success of the lycopods. There is very good evidence that the forests inhabited by *Lepidodendron* were swampy places. For a start, its remains are found in deposits of coal which, like peat, is the remains of plant material that has not decayed but merely been compressed over time. Second, there is evidence from the structure of the plants themselves; *Lepidodendron* had internal air channels (Fig. 6.4c) and spaces, just like modern aquatic plants (see Chapter 11) which would have improved the oxygen supply to submerged organs. Indeed,

the presence of air channels along the microphylls and stems suggests that they stood in water. Massive trees in swampy substrates have an anchorage problem. The solution evolved by lycopods was a two-component system of stem-like and root-like structures (Fig. 6.4b). Horizontal spread, providing a wide base for the system, was furnished by nearly horizontal stem-like structures called **rhizophores**. In turn these bore vascular root-like organs which came out at right angles all around the rhizophore, providing anchorage, as do the sinker roots of modern trees (Fig. 6.4d).

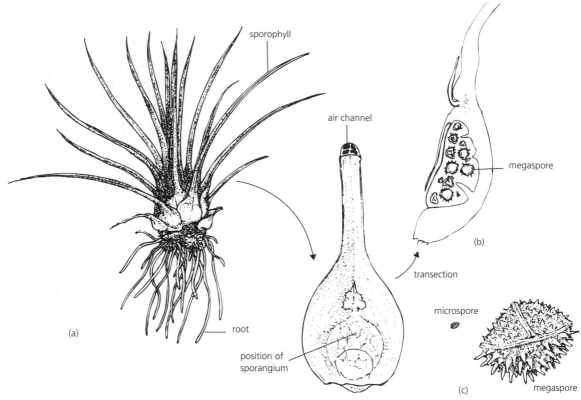

Fig. 6.5 The quillwort *Isoetes*, a modern survivor of the arborescent lycopods. (a) Side view (×1) with one sporophyll enlarged at the side to show the air channels that run through the leaves and an outline of the sporangium, which is well protected. (b) A section cut through a megasporangium showing the few large megaspores within. (c) A megaspore and a microspore together (×60) to illustrate their enormous difference in size. Quillworts show extreme **heterospory** (see section 6.6.4).

Modern lycophytes

The massive coal deposits of the world are testimony to the overwhelming success of the vegetative design of the arborescent lycopods. However, the only present-day survivors of these plants are a small and often overlooked group of plants called the quillworts, whose aquatic forms we will examine more closely in Chapter 11. Fossil evidence for the gradual reduction of a small arborescent lycopod to the small plants of *Isoetes* (Fig. 6.5) is good, and even the terrestrial forms of this genus have many of the features shown by arborescent fossils.

The herbaceous lycopods, the 'club mosses' which are alive today, such as *Lycopodium*, *Huperzia* (see Fig. 6.13, p. 103) and *Selaginella* (see Fig. 6.17, p. 109), share many of these features of the arbores-

cent forms, but are the direct descendants of herbaceous plants that are thought to have arisen alongside them. This does not mean that they are either small or short-lived, however. By scrambling over other plants many tropical species become very large. Arctic tundra species take root where soil is available and use their horizontal stems to travel across barren rocks to the next patch of soil. This latter strategy is known as guerrilla growth. In a nutrient-patchy situation such as arctic tundra this ability to travel horizontally allows *Lycopodium annotinum* plants to cover vast areas, and individual plants can undoubtedly live for several hundred years.

6.4.2 The horsetails (Sphenophyta)

Another group of plants that grew alongside the

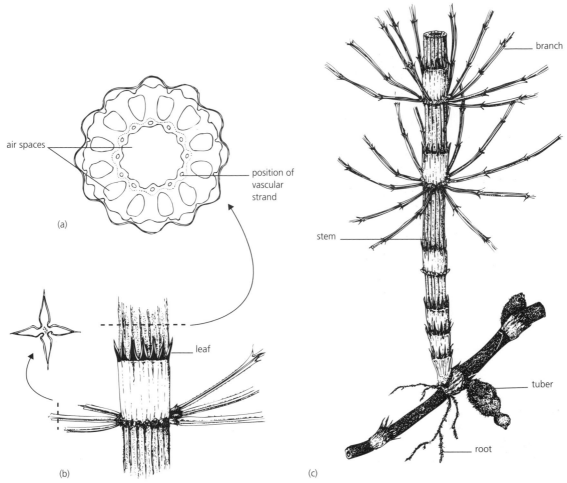

branch

air spaces

position of
vascular
strand

stem

(a)

leaf

tuber

(b)   (c)   root

Fig. 6.6 Morphology of the horsetail *Equisetum*. (a) Transverse section through the stem (×8), showing the peripheral and central air channels, with ring of vascular strands (**eustele**) round the latter. (b) Close-up of the stem (×2) to show the jointed construction. The topmost node bears a ring of tiny scale-like leaves, the lowermost bears a whorl of branches. Everything above the ground *but* the leaves shown here is photosynthetic! The transverse section through a branch alongside the drawing shows the four toughened points that form at the outermost edges which contribute to the effectiveness of this plant as a pot scourer. (c) Plant dug up towards the end of the summer (×1). The underground rhizome has formed tubers and bears roots at the internodes.

arborescent lycopods in Palaeozoic times, but that developed rather different methods for improving their competition for light, were the horsetails or sphenophytes.

Xylem arrangement

Unlike the rhyniophytes and whisk fern, whose xylem is arranged in a central rod—a condition known as **protostely**—that of sphenophytes is dis-

tributed either in a ring or in separate strands towards the outside of the stem (Fig. 6.6a). These arrangements, known as **siphonostely** and **eustely**, have two advantages. First, material arranged peripherally provides more support than material in a central rod, so the sphenophytes could grow taller without needing secondary thickening. Second, peripheral xylem facilitates branching of the structure, because branches can be more easily plumbed into the xylem system. In sphenophytes the parts all emerge in orderly whorls, giving the plants a 'jointed' appearance which explains their alternative name of Arthrophyta. Simple leaves, branches and roots may all be produced and they all emerge in rings, giving distinct nodes (joints) and internodes (Fig. 6.6b). It is at these solid nodes where extension growth takes place in the **intercalary meristems**. Some of the Carboniferous sphenophytes, such as *Calamites* also developed secondary thickening and grew into substantial trees up to 40 m in height. Other plants remained herbaceous.

Modern sphenophytes

Just one genus of the sphenophytes survives today, *Equisetum*, but what a success story *Equisetum* is! Like its ancestors and the lycopods, horsetails have both horizontal and vertical components. The big difference between sphenophytes and lycophytes is that the horizontal component, called a **rhizome**, travels under the substrate surface, not over it. Although this means it is not photosynthetic, so does not contribute organic carbon, it allows plants to ramify laterally and send up vertical parts into every tiny gap in the vegetation above. Couple this with the tough construction and a surface covered in silica

(Figs 6.6b and 6.7) and you have a plant that can penetrate tarmac! This makes it a feared weed in developed countries, and an efficient colonizer of moist places all over the world. Indeed, because of their extensive system of air channels, horsetails are even better suited than lycopods to thrive in water-logged soils.

One more advantage in having an underground rhizome is that it helps the plants survive at unfavourable times of year. Whereas the lycopods do not shed their leaves or above ground parts during the winter, some *Equisetum* species die right back. Towards the end of the growing season, they chan-nel organic carbon stores down into their below-ground parts and form next year's shoots. As soon as spring arrives, the fuel stores—starch-filled rhizomes in some, tubers (not unlike potatoes) in others (Fig. 6.6c)—power the intercalary meristematic acitvity that heralds the explosive vertical growth of the new shoots.

The die-back trick is similar to that used by some of the **deciduous** flowering plants of temperate environments (Chapter 9). It allows these plants to sit-out unfavourable conditions. In contrast, their evergreen counterparts have to battle to make up the losses of the respiration they cannot avoid with what-ever photosynthesis they can manage.

Fig. 6.7 Scanning electron micrograph of the surface of *Equisetum* (×2000) showing the projecting silica nodules which act as a defence against herbivores.

6.4.3 The ferns (Pterophyta)

In amongst the lycophytes and sphenophytes of Carboniferous forests were ancestors of the largest group of seedless vascular plants alive today, the ferns, or pterophytes. Like the two groups we have already examined, ferns developed a characteristic suite of adaptations which improved their competitive ability.

Megaphylls

The most conspicuous feature of most ferns is their attractive foliage. Their flat, and often intricately divided leaves (fronds) make them valuable horticultural commodities, and very efficient harvesters of light. These leaves are often large, and are called **megaphylls**, but are not distinguished from microphylls on the basis of size alone. Rather than representing the end-product of the evolution of tiny lateral projections, it is thought that megaphylls were once branched stems. If the topmost portions of a rhyniophyte descendant were flattened (**planation**) and the gaps between the parts filled in by tissue (**webbing**) an early megaphyll would be the result. The complex venation of the first ferns, in contrast to the single vascular strand in lycopods, is evidence for this theory. Fern leaves have always had highly branched and interconnecting veins, just like the vascular system of a branched set of stems. The branching was no doubt facilitated by the siphonostelic distribution of xylem which ferns, like the sphenophytes, have also developed.

Roots

Another vegetative adaptation of the ferns was a well-developed root system, which has helped some to develop into deciduous plants and grow well in seasonal areas. However, one adaptation that true ferns failed to develop is secondary thickening. Consequently, most ferns have short inconspicuous stems. Tree ferns do exist now, as they did in the Carboniferous period, but their trunks are produced in a rather unusual way. They have very little conducting tissue and are largely a mesh of shed leaf bases and **adventitious** roots which emerge from the top of the stem and travel downwards towards the

soil. The resulting tree is slow-growing and vulnerable to frost damage.

Modern ferns

The combination of a good basic body plan, with great adaptability of their life history, has allowed ferns to colonize most areas of the world and most habitats. Although ferns seem most common in tropical forests, several can survive hot dry conditions, for example by using a deciduous strategy, while in temperate regions two alternative strategies may be used. Some shed their leaves during the winter (see Box 6.1); others keep their foliage the whole year round and so exploit the increasing temperature and light of spring before deciduous competitors have sent forth their leaves. A final group of ferns have even reverted to a wholly aquatic existence, as we shall see in Chapter 11.

By far the most impressive modern ferns are some, such as the extraordinary bracken *Pteridium aquilinum*, that have stems that can grow further than the biggest trees alive in the Carboniferous or the present day, but in a horizontal orientation (Fig. 6.8a in Box 6.1). Some of these have stems that creep over the substrate, whereas the rhizomes of bracken tunnel beneath the surface, sending up vertical growths at intervals (as works so well for *Equisetum*). Much smaller in overall size, but just as conspicuous, are the typical 'shuttlecock'-type ferns (Fig. 6.8b) which have an upright stem like a tree fern, but one that is so short as to be easily overlooked.

Ferns grow particularly well in woodland, where their broad fronds efficiently harvest the little light that penetrates the canopy, and a typical woodland has all the growth forms. Those with horizontal travelling stems can make good **epiphytes** in rainforests, cashing in on the woody investment of trees by stationing themselves high in the canopy. Enough water and minerals run down the trunks to fuel their requirements, and they receive far better light than their counterparts on the forest floor. This strategy is also used by *Platycerium*, the stag's horn fern which does not scramble, but literally clings on (see Fig. 8.4a, p. 140). Specially modified leaves called clasping fronds, behind which accumulate debris perfectly adequate as potting compost, allow the

BOX 6.1 BRACKEN: A SPORING SUCCESS

Bracken, *Pteridium aquilinum*, is probably the most successful vascular plant in the world. It is found on every continent except Antarctica, and is one of the most abundant plants in many regions of our planet. The credit for this fern reaching the status of 'worst weed' in British farming literature lies both with our ancestors and with some superbly efficient design features of the plant.

Our Neolithic forebears did bracken a big favour by converting what had been forest to open agricultural land. This helped bracken because, although it is capable of growth in the understorey of forests, unusually for a fern it thrives in broad daylight and can stand a surprising amount of water stress. Its armoury of powerful defensive chemicals, many of which are carcinogens, also means that cattle and sheep avoid it. This gives it a decisive competitive advantage over grasses and heather in heavily grazed upland areas, allowing it to spread and take over the land. It has been estimated that the plant covers an area of Britain the size of the county of Devon (around 10% of upland

areas) and as land is set aside or becomes less intensively managed, this invader continues its spread.

One of the main keys to its invasive power is its extensive underground network of rhizomes. These underground stems can penetrate under, or grow through, other plants (or even roads!) and send up above-ground fronds wherever competition permits (Figs 6.8 and 6.9). Because these fronds can grow to over 2 m in height, they extend above most other herbs and, when fully expanded, easily shade them out. As the fronds photosynthesize and channel sugars back down to the storage rhizome, the growing tips forge ahead in all directions and colonize new sites. This creates an underground version of the lycopod guerrilla strategy (see section 6.4.1) which can potentially continue growth indefinitely. Although one section of rhizome probably lives for fewer than 40 years, genetically identical individuals have been mapped over a kilometre. The spores that started such plants off probably germinated over 10 000 years ago; the rhizomes have been growing at their tips and dying off at their bases ever since.

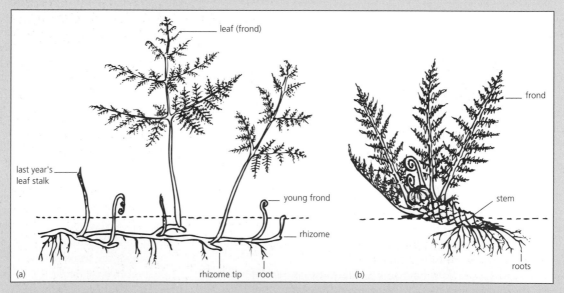

Fig. 6.8 Two of the growth forms of ferns. (a) Bracken (*Pteridium aquilinum*, see above), a fern with an underground rhizome bearing roots, the remains of last-season's leaves, buds and unfurling leaves, and fully expanded leaves. A bird's (or human's!) eye-view of such leaves gives no indication that they are connected under the surface. (b) *Dryopteris*, a fern with an 'upright' stem, which bears a crown of leaves that are more obviously connected to the same plant, in a shuttlecock-like arrangement (both ×0.05).

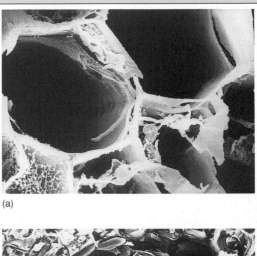

(a)

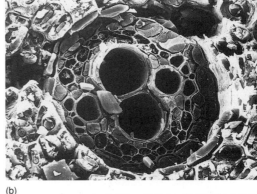

(b)

Fig. 6.10 Scanning electron micrographs showing an important vascular advance of ferns which include bracken: xylem vessels. (a) Xylem vessels in the stem (×200). (b) Vessels in the root (×60). These wide vessels transport water more efficiently than the thinner, more elongated water-conducting elements of many pteridophytes.

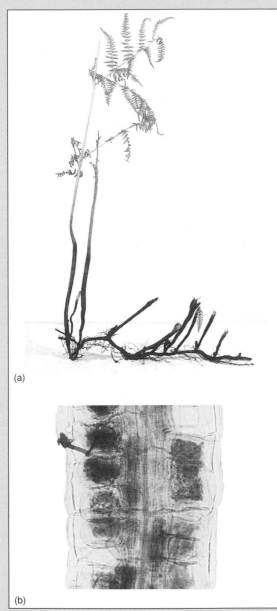

(a)

(b)

Fig. 6.9 Two important aspects of the morphology of bracken *Pteridium aquilinum*. (a) Growth form of an uprooted plant showing the underground rhizome. (b) Mycorrhizal (stained fungus) within a bracken root (×1000). In this mutualistic association the bracken gains improved phosphorus and water uptake in return for carbohydrates.

However, bracken does not rely exclusively on vegetative spread. In warm dry summers, each of the large leaves has been estimated to be capable of producing over 300 million spores. Dispersal is aided by the height of the fronds, because the spores are released above the boundary layer of air into currents that can transport the plant to islands over 3000 km from the nearest land mass.

A final key to bracken's success is its particularly efficient relationship with a mycorrhizal fungus (Fig. 6.9b). This greatly improves its uptake of nutrients, which enables it to thrive on the poorest and driest of

(*Continued on p. 100.*)

Box 6.1 contd.

soils, while water uptake is improved by its development of wide xylem vessels (Fig. 6.10). Almost the only natural phenomenon that deters bracken is late frosts which can kill the tender green fronds that have just emerged. But bracken has one trick left that ensures its survival, not only after late frosts, but also after mechanical or chemical attack from farmers. More buds lie dormant beneath the soil on the rhizome (Fig. 6.9a). If the first batch of fronds are killed, the plant releases a new batch, and will repeat the process until the reserves in the rhizomes are exhausted. The patience of many farmers is exhausted well before that!

plants to colonize even vertical tree trunks. The stag's horn-like photosynthetic fronds are then free to bask in the sun that penetrates the topmost parts of the forest canopy.

Most fern stems are well supplied with vascular tissue, which, as we mentioned before, is metabolically expensive stuff. Some specialists in the very dim recesses of rainforests and caves do not make the investment (Fig. 6.11). The 'filmy ferns' are so called because the blades of their leaves are only one cell thick, lacking stomata. Indeed, their economy of tissue is reflected even in their rhizomes, some of which have only a single xylem element running through them. Such plants have more in common ecologically with mosses than other ferns, being poikilohydric and heavily dependent on constantly moist conditions. The pay-off is that their lack of non-green tissues makes them exceptionally efficient survivors in low-light environments.

6.5 ASEXUAL REPRODUCTION OF VASCULAR PLANTS

Asexual reproduction is much more common in land plants than in seaweeds, perhaps because large propagules do not risk being swept so far away from somewhere suitable to set up home. The ferns and their allies have an enormous range of strategies by which they reproduce asexually.

6.5.1 Simple mechanisms

The simplest form of reproduction from vegetative parts is fragmentation, and most ferns and fern allies grow well from fragments. This is particularly true of the horizontal-growing forms, such as *Equisetum*

and *Pteridium* (see Fig. 6.6 and Box 6.1), accounting for much of their persistence as weeds, but some upright forms also grow well from fragments. Growers of some tree ferns regularly cash in on this by cutting off the topmost parts and placing them back on soil (thus getting a 'new' plant when the old one would have had to be removed for the sake of the glasshouse roof!). Because they have **adventitious** roots (which come straight out of the sides of the stem), they can quickly re-establish water and nutrient conduction, just as they do in natural rainforest situations when the plants topple over.

6.5.2 Stolons

Stolons are one specialized means of reproduction familiar to anyone who has seen or grown strawberry plants, the slender stem-like organs that issue from ferns such as *Nephrolepis* (Fig. 6.12a) and the tropical club moss *Lycopodiella cernua*. Like stolons of strawberries, these can generate new plants when they touch the soil. The same is true of the leaves of many ferns, the most famous of which is the walking fern, *Asplenium rhizophyllum* (Fig. 6.12b) whose leaves, as its name suggests, can form roots and start off new plants.

6.5.3 Bulbils

Perhaps the most ingenious method is to produce tiny replicas of the parent plant, called **bulbils**. Many familiar house plants such as *Kalanchoe* and mother ferns manage to cover substantial areas of their natural habitat using this method. Mother ferns form miniature replicas of themselves on their leaves which, when there are enough of them to weigh the leaf down, extend roots down into the soil and estab-

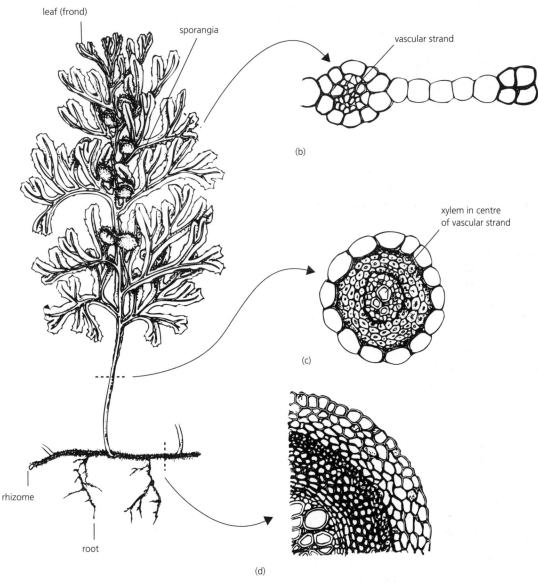

Fig. 6.11 Tunbridge filmy fern *Hymenophyllum tunbrigense*. (a) View of plant (×2), which is constructed on similar lines to those of *Pteridium*; the fern has both horizontal and upright portions, like bracken, but with very different anatomy. The blades of the leaves (b) are almost transparent, being only a single cell thick except in the centre and at the extreme edges, which are slightly thickened. Sections through the petiole (c) and rhizome (d) show the tiny amount of conducting tissue arranged in a **protostele**.

lish new plants. Lycopods such as *Huperzia selago* (Fig. 6.13) go one better and construct bulbil bases that cause the tiny replicas to leap well clear of the parent plant when they are disturbed by the wind or by an animal brushing against them.

6.6 SEXUAL REPRODUCTION OF SPORE-PRODUCING PLANTS

The life cycles of most of the ferns and their allies are broadly similar to that of *Psilotum*. The sporophyte

101

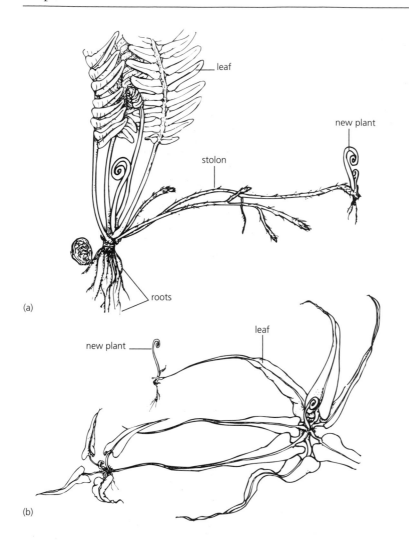

leaf

new plant

stolon

roots

(a)

new plant

leaf

(b)

Fig. 6.12 Vegetative reproduction in ferns. (a) *Nephrolepis*, a popular house fern which produces hairy stolons capable of generating new plants when they reach a patch of soil. (b) Walking fern (*Asplenium rhizophyllum*) 'walks' around the environment by producing new plants at the tips of its leaves (both diagrams ×0.25).

produces desiccation-resistant spores which germinate in moist conditions to produce gametophytes independent of the sporophyte. In turn these produce motile male gametes which fuse with eggs protected within an archegonium and grow into a new sporophyte. However, some plants show advances that improve the efficiency of this process, and one such development, **heterospory**, was a crucial step on the road to the evolution of seeds.

6.6.1 Production and dispersal of spores

The earliest land plants, as we have seen, bore sporangia at their tips, where they are vulnerable during their development. One thing that distinguished the ancestors of lycopod, such as *Asteroxylon*, was that their laterally positioned sporangia developed amongst, and therefore protected by, microphylls. The arrangement in some lycopods is still more sophisticated, and their sporangia are borne on the top surface of microphylls, often bearing extensions of protective tissue; in these cases they are referred to as **sporophylls** (see Fig. 6.5). Plants whose sporophylls are aggregated together are said to have **strobili** or cones (the 'clubs' of some club mosses—Fig. 6.13).

Sphenophytes have cones whose origins are not thought to lie with modified leaves, but with stems. If the tips of plants with terminal sporangia were

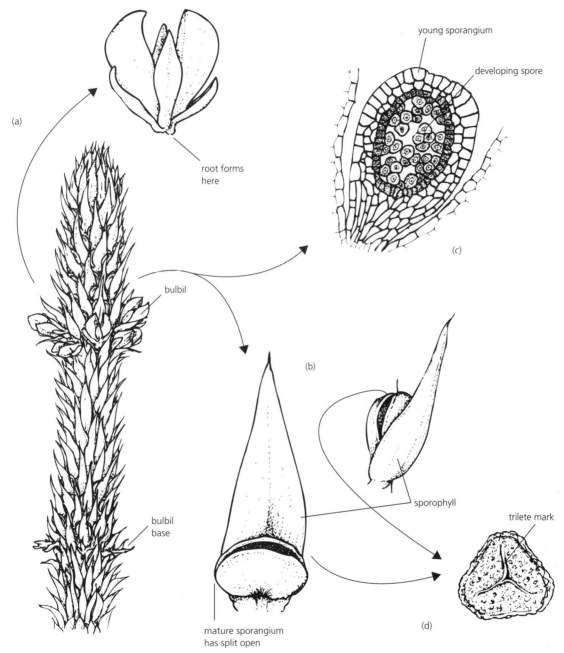

young sporangium

developing spore

(a)

root forms here

bulbil

(c)

(b)

bulbil base

sporophyll

trilete mark

mature sporangium has split open

(d)

Fig. 6.13 Reproductive structures of the club moss *Huperzia*. (a) The tip of a stem (×3) which generates miniature plants, bulbils, in a regular fashion. The bases of those generated last year can be seen at the base of this stem, giving a marker for a year's growth rather like tree rings. Between each round of bulbil production comes a round of sporophyll production. A single sporophyll is enlarged in side and bird's eye-view to the side (b), showing how the developing sporangium (c) is protected until mature and ready to release the spores (d) (×400).

curved over, a structure resembling the **sporangio-phores** of *Equisetum* (Fig. 6.14) would be the result. The cones of sphenophytes, which superficially resemble those of lycopods, are therefore thought to be aggregations of side shoots bearing terminal sporangia.

The spores of ferns, like those of lycophytes, are generally associated with leaves, but do not form cones, instead being held around the edge or under the lower surface of the leaves. Here they are protected by special umbrella-like coverings or flaps of tissue called **indusia**.

Whatever their position on the plant, sporangia rely on the same broad categories of dispersal vectors as the bryophytes—wind, water and animals—and employ similar mechanisms to ensure that their spores come into contact with these agencies. The simplest sporangia, such as those of *Huperzia* (Fig. 6.13b, c), just split open and rely upon wind or passing animals to carry the spores away. Some of the most sophisticated sporangia are found in ferns like *Pteridium*—they function like miniature sling shots, flinging the spores as far as possible into the air currents around the leaves (Fig. 6.15). Although the vast majority nevertheless land very close to the parent plant, many are subsequently washed away by rainfall and some make it right up into air currents responsible for carrying them up to thousands of kilometres away.

The spores of most species are tiny, tough, thick-walled bodies, able to survive long-distance dry air travel. Many years may elapse before they encounter the combination of factors they require for germination, however. In most cases ideal conditions are light and damp with adequate nutrients and few competitors. In the meantime they may form part of a spore bank in the soil, remaining dormant until returned to the surface. At the other extreme are spores that are thin-walled and green, like those of *Equisetum*, the royal fern *Osmunda* and the filmy ferns. These are short-lived but ready to start growing the moment they hit the ground.

6.6.2 Growth of the gametophytes

Some lycopods and ferns have subterranean gameto-phytes comparable to those of *Psilotum*. Their

metabolic needs are partly serviced by helpful fungi, and their male gametes manage to find their way through the soil to fertilize the eggs of other plants. The majority of gametophytes, however, are photosynthetic, surviving on the surface of substrates like newly exposed mud or earthworm casts. Most gametophytes mature quickly, producing sporophytes just a few weeks after spore germination. If the new en-vironment colonized is suitable, these sporophytes may be the start of hundreds or even thousands of years of growth. This potential longevity may be one reason why gametophytes are the less familiar part of the life cycles of pteridophytes such as *Equisetum*, *Lycopodium* and *Pteridium*.

Once they arrive at a substrate and encounter the right combination of light, warmth and moisture, most spores germinate rapidly. Most generate a filament of photosynthetic cells which is anchored and supplied with moisture by simple unicellular rhizoids (Fig. 6.16). They may lack the thick waxy cuticles and protective structures of their sporophytes but despite their apparent vulnerability, the gametophytes of some species are more tolerant of environmental extremes than sporophytes. Once the tip of the filament detects sufficient light, cell division generates a new cell wall parallel to the direction of growth, rather than at right angles, and the filaments start to produce two-dimensional growth. The final form of many fern gametophytes is a heart-shaped plate of cells (Fig. 6.16h); although those of *Equisetum* and some lycopods are chunkier and more three-dimensional.

6.6.3 Reproduction of gametophytes

Fertilization between two gametophytes might appear to be a real problem. After all, the chances of two gametophytes germinating at similar times and close enough to allow the sperm to swim between them must be low. Most are capable of producing both male and female gametes simultaneously and self-fertilizing if they fail, but they have two neat ways of improving their chances of cross-fertilization.

The first feature that enhances the chances of outbreeding is a separation of sex organ formation. This may be temporal separation—usually it is female

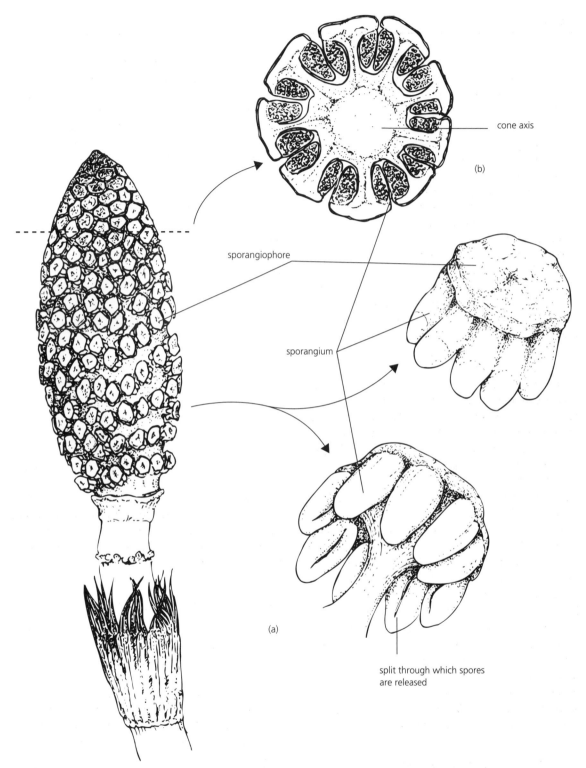

cone axis

(b)

sporangiophore

sporangium

(a)

split through which spores
are released

Fig. 6.14 Reproductive structures in *Equisetum*. (a) A fertile stem bearing a cone (×3). One sporangiophore is shown in detail to the right. As in *Huperzia*, the developing sporangia are protected until they are mature, but in this plant that is achieved by the close fit between the sporangiophore tops (look at the cone apex and the section cut through (b) and the reflexion of the sporangia).

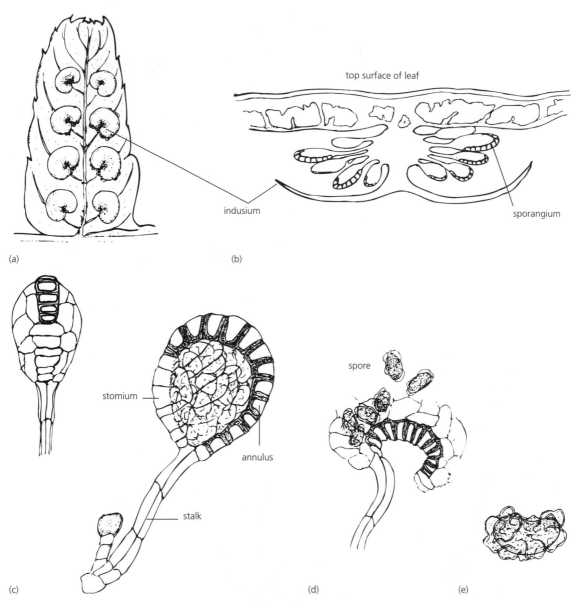

Fig. 6.15 Reproductive structures in ferns. (a) The underside of a pinnule of a *Dryopteris* leaf, bearing the kidney-shaped protective structures (indusia) characteristic of this species. (b) Section cut through a leaf, revealing that the indusia protect developing sporangia. Once mature, the elaborate heads of the sporangia poke out into the dry air and the band of empty thickened cells that runs around all but one part of the sporangium (the annulus) (c) (×400) starts to dry out. Finally the tension on the weaker, thin-walled cells becomes too great and the whole thing rips open (d), the annulus then flings the spores (e) (×1000) into the dry air with a slingshot action.

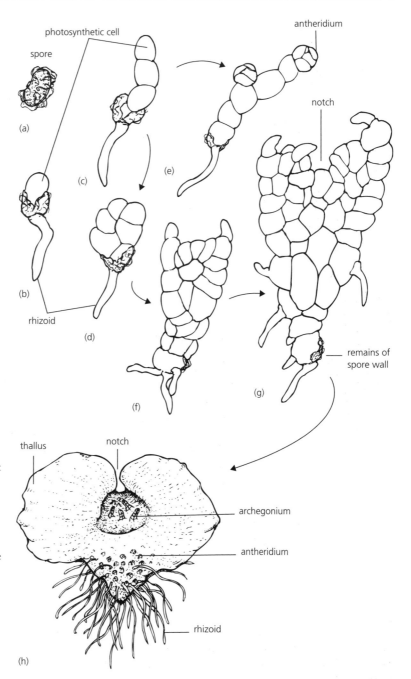

Fig. 6.16 Germination and growth of a fern gametophyte. (a) A spore of *Dryopteris* (×500) which has landed somewhere suitable germinates, putting out first a rhizoid then a photosynthetic cell (b). A short filament of cells (c) usually continues to produce photosynthetic cells (d) but under the influence of antheridiogens, can form antheridia (e). The latter remain male; the former broaden out to form a one-cell thick plate-like thallus anchored by rhizoids (f, g), the central portion of which becomes more than one cell thick and develops archegonia just behind the notch (h) (×25). Such archegoniate gametophytes also produce antheridia in some ferns, as shown amongst the rhizoids, but most are exclusively male or female at any one time.

gametangia that form first (we will see why shortly), and antheridia only appear after the archegonia have matured. If the two sorts of gametangia are produced at the same time they are usually spatially separated; the archegonia are produced towards the notched region of the gametophyte, and may have their necks turned away from the antheridia, positioned at the other extremity (Fig. 6.16h).

The second feature that can vastly improve the chance of a gametophyte recruiting sperm from a different gametophyte is their production of extremely potent chemicals called **antheridiogens**. These are generated by gametophytes that have matured sufficiently to start producing archegonia, and they are capable of replacing the light requirement for spore germination. This means that spores lying just beneath the surface of the soil, or under leaves near the gametophyte in question, are prompted to germinate. They may not have sufficient light to grow very much, but here a second effect of the molecule becomes crucial—they are prompted to form antheridia rather than photosynthetic cells. The antheridia in turn rapidly produce sperm which fertilize the original gametophyte. Antheridiogens are capable of exerting their influence at incredibly low concentrations (10^{-17} M!) and at their most effective can result in a gametophyte which consists only of a single antheridium. So, despite their similar genetic make-up, spores can germinate to produce two very different types of gametophyte.

6.6.4 Heterospory

In most of the plants we have described in detail so far, the sporangia all look alike and produce spores that look identical and, given the same opportunities, behave similarly. These plants are described as **homosporous**. As we have just seen, though, with such an arrangement cross-fertilization is not guaranteed.

Some of the plants in each of the groups we have examined have continued an important evolutionary trend that we noted first in algae such as *Fucus* (see Chapter 4). This is the increasing provision of nutrients and protection for structures destined for the female line, balanced by economy and streamlining on the male side. *Fucus* is oogamous, having large,

well-nourished female gametes, housed in protective structures. The motile male gametes are much smaller, but produced in greater numbers than female gametes. In some lycophytes, sphenophytes and pterophytes these trends were taken one stage further with the development of **heterospory**. In heterosporous plants, it is not only the male and female gametes that differ from each other, but the spores that generate the gametophytes from which the gametes form.

Some arborescent lycopods and calamites were heterosporous, as are the quillworts *Isoetes*, the spike mosses of the genus *Selaginella*, and the water ferns. A close look at the cone of a *Selaginella* plant reveals two sorts of sporangia (Fig. 6.17). The smaller sporangia contain numerous tiny spores, called **microspores**; the larger sporangia have just four much bigger spores, called **megaspores** (see also Fig. 6.5). The micro-spores will go on to generate gametophytes which are exclusively male, the megaspores will produce female gametophytes.

6.6.5 Reduction in gametophyte size

Both the male and female gametophytes of *Selaginella* also demonstrate a second tendency, common in plants showing heterospory–a reduction in size from the free-living homosporous condition. Not only are they much smaller than in most homosporous plants, but they are **endosporic**, never really leaving the protection of their spore coats, and rather than photosynthesizing) they live off stored food. They generally produce very few cells–all their efforts are concentrated into the production of gametangia. In *Selaginella* (Fig. 6.18a) the megaspores do generate somatic cells, but the only parts of the gametophyte that protrude from the protective megaspore wall are a few rhizoids (which do not reach the substrate but probably aid water retention) and the archegonia. The microspores generate only spermatogenous tissue, all of which is converted to sperm. This whole process improves the likelihood of sexual reproduction involving cross-fertilization. The small gametophytes mature quickly; they are likely to survive within the protective spore coat; and they do not have stringent requirements for germination.

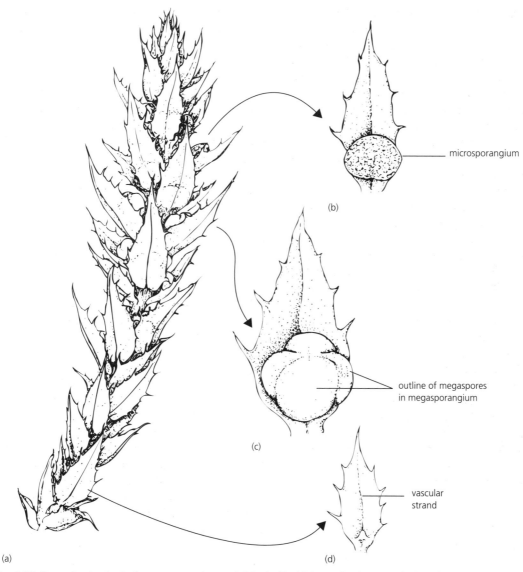

Fig. 6.17 Reproduction in the heterosporous lycopod *Selaginella*. (a) Reproductive cone (×4), with two sporophylls and a microphyll dissected out. The topmost sporophyll (b) bears a microsporangium containing many small microspores; the lowermost sporophyll (c) bears a megasporangium within which are four large megaspores. (d) Microphyll.

6.6.6 Retention of the megaspores

The efficiency of reproduction is further increased in some species of *Selaginella* by retaining the megaspores on the parent plant. Some arborescent lycopods enveloped the megasporangium in an extra layer of sporophyll. These measures provide even more protection and a more controlled environment for the gametophyte. Both also call to mind the features of the seed plants we will examine in the next chapter. Indeed, although true seeds were developed only by a single group of plants, it seems that heterospory arose independently in several groups, each of which went some way towards producing seed-like structures.

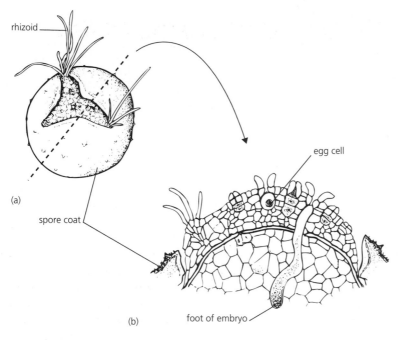

rhizoid

(a)

spore coat

egg cell

(b)

foot of embryo

Fig. 6.18 Reproduction in the heterosporous lycopod *Selaginella*. (a) A germinated megaspore of *Selaginella* (×100). This mature gametophyte is still protected by the spore wall; apart from the few rhizoids the only parts exposed to the outside world are those bearing the archegonia. (b) A section through the same plant; one egg has been fertilized and the embryo has sent down a 'foot' to take nutrients from the parent gametophyte while it builds up enough reserves to break out of the protection of the original spore coat.

6.7 POINTS FOR DISCUSSION

1 We know that the leaves of some arborescent ancestors of modern-day horsetails were large, branched structures with branched veins—do you think the tiny scale-like leaves of *Equisetum* (see Fig. 6.6) should be called microphylls or megaphylls?

2 How do you think Stag's horn ferns get into the tops of trees?

3 Spores can be produced in far greater numbers than seeds, so why is the earth not covered with bryophytes and ferns?

4 What are the potential disadvantages of heterospory?

FURTHER READING

Camus, J.M., Jermy, A.C. & Thomas, B.A. (1991) *A World of Ferns*. Natural History Museum Publications, London.

Cleal, J.C. & Thomas, B.A. (1999) *Plant Fossils. The History of Land Vegetation*. Boydell Press, Suffolk.

Kenrick, P. & Crane, P. (1998) *The Origin and early Diversification of Land Plants*. Smithsonian Institution Press, Washington, US.

Stewart, W.N. & Rothwell, G.W. (1993) *Palaeobotany and the Evolution of Plants*, 2nd edn. Cambridge University Press, Cambridge.

(a)

(b)

(c)

(d)

Plate 1 Surviving in difficult environments I. Many mosses and ferns can survive in intermittently dry environments like deserts or stone walls due to their poikilohydry; they can dry out and shut down their metabolism wihout being damaged. (a) The moss *Hedwigia integrifolia* from the mountains of Cameroon in its dry state. (b) The same plant revived after rewetting. (c) The rustyback fern *Ceterach officinarum* on a dry wall looking dead and shrivelled. (d) Ferns on the same wall after a rainshower. The rustyback has rehydrated and is the fern with the larger once-divided fronds.

[facing p. 110]

Plate 2 Surviving in difficult environments II. Plants use many different ways of surviving on or under the water. The fern *Salvinia auriculata* (a) floats on the water surface obtaining nutrients through its root-like leaves. Buoyancy is provided by densely packed egg-beater trichomes (b) on the uppper surface of its fronds which repel water. In contrast, in the Amazon lily *Victoria amazonica* (c) only the large circular leaves float, and oxygen is piped down to a conventional root system. The sea palm *Postelsia palmaeformis* (d) can cope with both submersion and (as here) exposure on the wave-battered Californian coast, being glued to the rocks via a complex branched holdfast.

(a)

(b)

(c)

(d)

Plate 3 Reproduction and dispersal I. Spore-producing plants use many different techniques to power dispersal of their propagules. Many mosses such as *Polytrichum piliferum* (a) develop antheridia within funnel-shaped cups. Raindrops landing in the cup cause splash dispersal of the sperm. A similar mechanism (b) is used by lichens such as *Cladonia chlorophaea* which use splash cups to disperse their soredia. Most other spores are distributed in dry conditions and are dispersed by the wind. In the sporophyte of the moss *Rhynchostegium confertum* (c) desiccation causes the teeth at the end of the capsule to open and release the spores into the airstream. In the lichen *Cladonia floerkiana* (d) desiccation promotes release of fungal spores from the red hat-like structures.

(a)   (b)

(c)   (d)   (e)

Plate 4 Reproduction and dispersal II. Angiosperms use many different animal partners to pollinate their flowers, which each have a characteristic morphology. Bat-pollinated flowers such as those of the balsa tree *Ochroma lagopus* (a, b) are stout and colourless with a heady scent and large nectar reward. In (a) a bat is feeding, in (b) flying away. The cape heath, *Erica versicolor* (c) is a typical bird-pollinated flower with tubular red flowers and no scent. *Buddleia* (d) is a typical butterfly-pollinated flowers, having tubular purple scented flowers. Here it is being visited by the long-tongued hummingbird hawkmoth, which is hovering in front of the inflorescence. The fly orchid *Ophrys insecta* (e) like many other orchids is a mimic. It is pollinated by male bees which mistake the flower for a female tree and attempt to mate.

(a)

(b)

(c)

(d)

Plate 5 Reproduction and dispersal III. More ways angiosperms use to achieve pollination. Most angiosperms, like the mayweed *Tripleurospermum inodorum* (a), provide nectar and pollen rewards for visitors and attract them with conspicuous flowers. This daisy-like 'flower' is, in fact an inflorescence made up of large numbers of yellow tubular flowers within a disc of sterile white strap-like flowers. Flowers like these are popular with worker bees. A few flowers, such as the arum *Dracunculus muscivorus* (b) provide a heavy scent of rotting flesh. The flower looks like a rotting seagull and is pollinated by carrion flies which mistake it for the sites on which they lay their eggs. Some common angiosperms have reverted to wind pollination. These include the grasses such as *Arrhenatherium elatius* (c) which release pollen from their exposed anthers and catch pollen on their feathery stigmas. Many temperate trees are also wind pollinated. In the hazel *Corylus avellana*, wind-borne pollen blown from the male catkins is caught on the exposed red stigmas of the female flower (d).

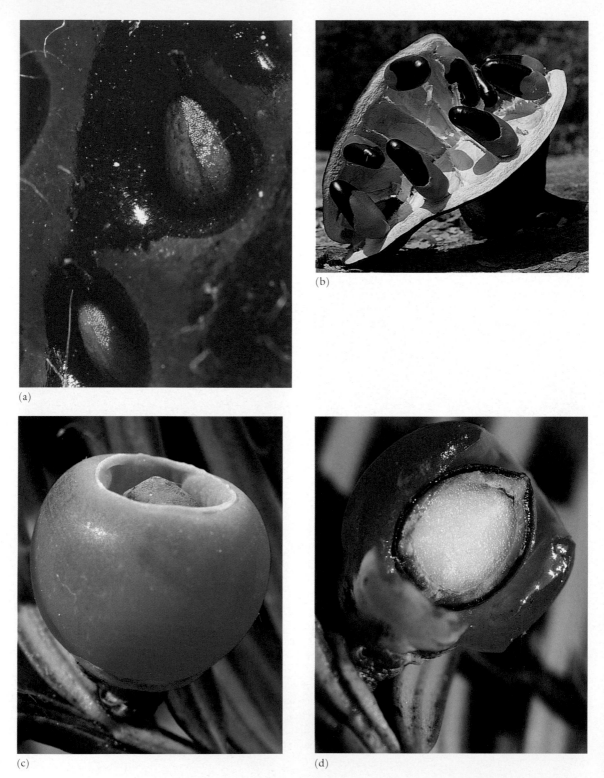

Plate 6 Reproduction and disperal IV. Many seeds are dispersed by birds, which are attracted by a reward of red fruit. In the strawberry *Fragaria vesca* (a) a single large fruit is coated with many tiny seeds. However, some plants exploit birds' attraction to the colour red. In the legume *Afzelia bella* (b) the seeds mimic fruit by being red, but they have no nutritive value. They simply pass through the gut of a bird which eats them unharmed. The yew *Taxus baccata* produces what looks like red fruits (c). Examination of a longitudinal section (d), however, reveal them to be arils, outgrowths of the seed coat.

(a)

(b)

(c)

(d)

Plate 7 Exploitation of other plants. The strangling fig *Ficus* (a) exploits the support of another tree to reach the light, before wrapping its roots around the trunk and killing it. The hemiparasite *Castilleja* (b) from Utah takes more, tapping into the xylem stream of its neighbours' roots for water and nutrient salts. However, it still carries out photosynthesis. The holoparasite *Rafflesia pricei* (c) taps into the phloem stream of lianas, extracting everything it needs. The only visible structure is the 60 cm diameter flower which attracts flies by mimicking the smell of a rotting corpse. The yellow birdsnest *Monotropa hypopitys* (d) is a saprophyte, living on decaying organic matter and like *Rafflesia* having no need for chloroplasts.

(a)

(b)　　　　　　　　　(c)　　　　　　　　　(d)

Plate 8 Exploitation of insects. Many species of plants have mutualistic relationships with ants, providing them with a home in return for protection from herbivores and climbers. The tropical tree *Barteria fistulosa* (a) has hollow twigs in which the ants make their nest. In *Dischidia pectenoides* the ants are provided with a nestbox (b) made from a modified leaf. Cutting this open (c) reveals a clump of roots which absorbs nutrients from the ants' droppings. In contrast to these mutualistic associations the sundew *Drosera anglica* (d) is carnivorous, trapping ants and other insects on its sticky hairs and slowly digesting and absorbing their bodies.

Life from seeds

7.1 INTRODUCTION

Some of the heterosporous plants we considered in the last chapter not only retain their megaspores on the parent plant but may also protect them by encasing their sporangia. This extra care for the up-coming generation is thought to have given such plants an advantage over plants whose spores are freely dispersed, and whose embryos are at the mercy of the elements. Many groups of plants showed such developments during the Devonian period, including the progymnosperm *Archaeopteris* (Fig. 7.1), a tree which grew to heights of up to 20 m. However, further development led to the eventual production of **seeds** and the rise of the seed plants.

7.2 THE STRUCTURE OF SEEDS

The process of seed production begins with an **ovule**. This is a megaspore which is retained not only within a megasporangium but within an extra layer or two of tissue. The extra layers, the **integuments**, may have been derived from leaf parts or from axes, but these have since been reduced to simple envelopes with just a small opening to allow fertilization. The ovule is fertilized by sperm from a microgametophyte, which, in seed plants is inside the **pollen grain**. Once fertilized, the structure turns into a **seed** and the integuments develop into a **seed coat**, a layer which seals around the whole structure and protects the embryo. Seeds contain stored nutrients, just like the megagametophytes of free-sporing heterosporous plants. However, the advantage enjoyed by the embryo of a seed plant is that it can tap into these nutrients and develop all its parts while still tucked safely inside the seed coat.

7.3 THE EARLIEST SEED PLANTS

The oldest seeds identified so far are from the late Devonian, about 360 million years ago. The most notable feature of these seeds was their primitive integuments, which extended beyond the apex of the ovule. The plants that produced these seeds were a group with foliage very like that of modern ferns, the seed ferns. A typical member of this group was *Medullosa* which looked superficially very like a tree fern. The seeds produced by the seed ferns were little different from structures found in connection with arborescent lycopods. The small differences proved critical, however, and the Carboniferous explosion of spore-producing plants was followed by an explosion of seed-producing plants very much at the expense of their free-sporing relatives. This replacement was no doubt speeded up by climatic change. Seed plants would have had increased competitive advantage over free-sporing plants in the much drier Permian period.

In their turn, the seed ferns gave way to a number of different groups of plants which demonstrated further evolutionary advances. Many showed changes in their vegetative structure which were similar to those made by the spore-producing plants, and most groups also show further advances in reproduction. However, the advantage conferred by seed production has ensured that seed plants have dominated the land for over 250 million years. There are three main surviving groups: the cycads, the conifers and the angiosperms.

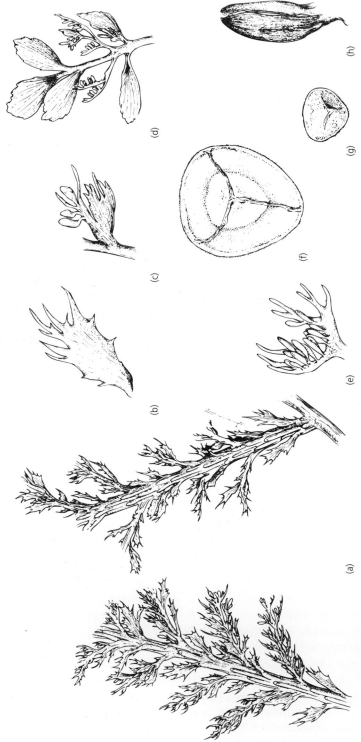

Fig. 7.1 *Archaeopteris*, a progymnosperm. These drawings have been compiled from reconstructions and fossils of the plant and show the foliage. Palaeobotanists suggest that the structures coming off the main stem shown here are the equivalent of leaves and small branches bearing leaves (a) (×0.2). Some leaves (b) were sterile, while others (c) bore reproductive structures. The terminal portion of a branch shown in (d) demonstrates that some branchlets had negligible leafy tissue, and were covered in sporangia (e). (f) A megaspore, alongside which is a microspore (g) (both ×100). The division between fertile and sterile parts of these plants did not seem to be as clear cut as modern plants, and leafy enclosure of sporangia (h) such as those of these progymnosperms is thought to have been what generated the seeds of the first seed plants.

7.4 THE CYCADS (CYCADOPHYTA)

7.4.1 Vegetative structures

The cycads are an ancient group of seed plants which, together with the extinct group the bennettitales, and the ginkgos and conifers that we shall meet next, dominated the world during the relatively dry Permian and Triassic periods. Cycads are known to have been a favourite food of some dinosaurs during the later Jurassic and Cretaceous, but they have since declined and only about 140 species survive, restricted to the hotter areas of the world. In their vegetative form, cycads resemble tree ferns. Like tree ferns they have simple crowns, with little branching, while the trunks of arborescent cycads are supported by persistent woody leaf bases (Fig. 7.2) just like tree ferns. This rather prescribed pattern of growth is no doubt a major reason for their current lack of diversity.

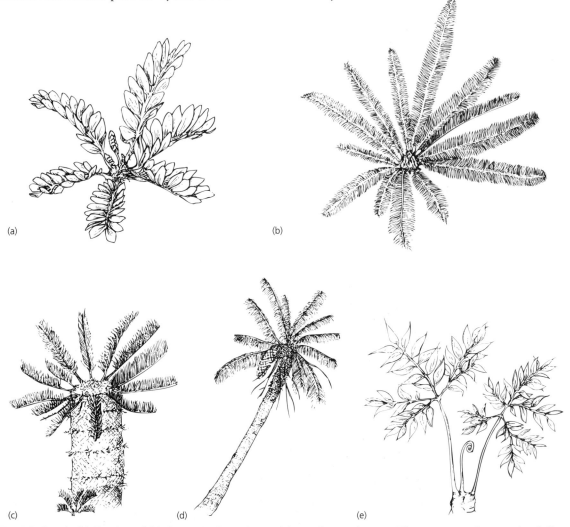

Fig. 7.2 Cycads. (a) *Zamia*, and (b) *Dioon*, (both ×0.1) two of the smaller cycad genera. The stems are underground and all that appears above the soil is the crown of leaves and the cones. (c) *Encephalartos*, (d) *Microcycas* and (e) *Bowenia* (all ×0.015) are all genera that include massive trees, support for which comes from the woody bases left behind when each crown of leaves dies back.

113

7.4.2 Reproduction

Cycads show big advances in their method of sexual reproduction. The reproductive units themselves are mostly cone-like structures, with reduced leaves enclosing attached sporangia. The real advances are in the fine detail of fertilization. When the pollen grains land on the female cone and germinate, they do not just release their sperm into the surrounding fluid as heterosporous plants do. Instead the pollen grains produce a **pollen tube** which grows as far as the opening to the ovule (the **micropyle**). The motile sperm can then swim down the tube, and when they reach the end are ideally placed to fertilize the ovule. Once more, though, the primitive nature of cycads is revealed because the male sperm still have flagella. The sperm of *Zamia* (Fig. 7.3) are the largest in the world, at over 0.5 mm, and are powered by large numbers of flagella.

Fertilization of the female generates seeds which are protected by the enormous cones in all but one, thought to be the most primitive, of the living cycads (Fig. 7.4). In the exception—the false Sago palm *Cycas*—the seeds are protected by a range of defensive chemicals. It was these seeds that were responsible for poisoning members of Captain Cook's crew when they made flour from the seeds, but omitted some vital steps used by Polynesians to leach out the toxins.

Many cycads are much prized horticultural favourites, and are among the few plants for which there is an artificial insemination scheme! One reason why a pollen store is necessary is that the plants are either male or female—they are **dioecious**. Their rarity, added to the infrequent production of reproductive structures means that female plants may have little chance of becoming fertilized by natural means. However, some cycads do use insects as pollination vectors, attracting them, like the arum lilies we will examine in Chapter 8, by producing excess heat.

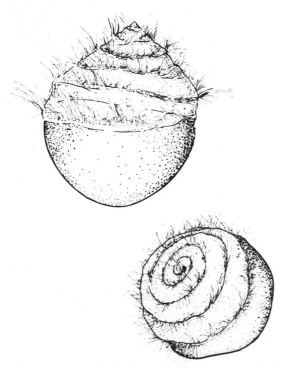

Fig. 7.3 Two views of *Zamia* sperm (×80). These enormous cells are not streamlined. They need swim only a very small distance to the egg cell of the female as the pollen is delivered to the female cones along a pollen tube. They are propelled by the many flagella that spiral around almost half the cell.

7.5 THE EVOLUTION OF SECONDARY THICKENING

The seed plants all share a further advance in their vegetative structure whose importance rivals that of the evolution of seeds. This was the development of a layer of a ring of cells around the outside of their stems which remains capable of cell division both to the inside and outside of the stem. Such a **vascular cambium** confers the ability to produce true **secondary thickening**.

Substantial secondary xylem, which has the dual role of conducting water up the plant and strengthening it, is laid down on the inside of the cambium. Secondary phloem, which has the role of transporting food down the plant, is laid down to the outside. This important innovation allows plants to grow thicker as they get taller and facilitated the evolution of enormous trees. It also allows the growth of a large, well-branched canopy which can effectively harvest light. Trees with fully developed secondary

Fig. 7.4 Reproductive parts of female cycads. (a) shows a single megasporophyll of *Cycas*, the only genus lacking distinct female cones. This structure is somewhat similar to some of those of *Archaeopteris* (Fig. 7.1), in that there is fertile and sterile tissue on the same structure. The reproductive parts here, however, are seeds. All the other female cycads have cones, some of which are huge and heavy. Their production is so metabolically 'expensive' that these plants do not produce leaves while they are generating cones. (b) *Stangeria*, (c) *Dioon*, (d) *Zamia*, (e) *Encephalartos*, (f) *Macrozamia*, (g) *Lepidozamia*.

thickening could therefore outcompete neighbouring tree ferns, cycads and seed ferns.

Cycads never seem to have developed the ability to lay down a great deal of secondary xylem. Their secondary growth is only sluggish. Instead, the first trees to exhibit rapid secondary growth were spore producers, which were successful formers of forests in some regions during the late Devonian period. These plants, progymnosperms like *Archaeopteris*, are thought to have given rise to the two main groups of plants alive today, the conifers and the angiosperms.

One further group, the **Ginkgophyta**, is represented by a single species, the maidenhair tree *Ginkgo biloba* (Fig. 7.5). This living fossil has changed little for over 150 million years but the ginkgos were already being outcompeted in the Permian by other groups, including the conifers.

7.6 THE CONIFERS (CONIFEROPHYTA)

The conifers are one of the most successful groups of plants today; conifer forests cover wide swathes of the temperate and subpolar regions. The features that account for their evolutionary success include both vegetative and reproductive characters.

7.6.1 Vegetative structures

Obviously, rapid secondary thickening gave conifers a big advantage, allowing them to compete with other plants for light. But because the secondary xylem of conifers is made up almost entirely of cells of a single type, **tracheids** (see Fig. 7.8 below), they show some compromise in their design. Tracheids

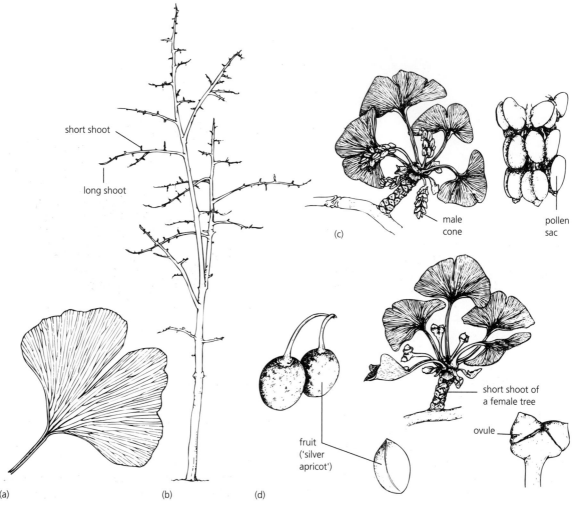

short shoot

long shoot

(c)

male cone

pollen sac

short shoot of a female tree

ovule

fruit ('silver apricot')

(a)   (b)   (d)

Fig. 7.5 *Ginkgo biloba*, the maidenhair tree. The specific name comes from the attractive leaves, which have two lobes (a) (×1.0). Unlike many gymnosperms this tree is deciduous and, the winter outline of a young specimen is shown in (b) (×0.02). This species is widely cultivated, and surprisingly pollution-tolerant for a 'living fossil', but the trees that line streets in New York are all males (c) as the fleshy exterior of the seeds produced by female trees (d) (×1.0) have an extremely unattractive smell!

are long and thin, so the wood they produce provides excellent mechanical support to the plant. However, being narrow, the tracheids are rather poorer than vessels (see section 7.7.1) at conducting water. Despite this disadvantage, conifers are the largest living organisms on the planet. Large specimens of the giant redwood *Sequoiadendron giganteum* from the western United States weigh hundreds of tonnes, many times more than a blue whale!

Secondary thickening is not the only important

vegetative feature of conifers. Many modern species possess narrow needle-like leaves, but this is not an original feature since many of the earliest conifers had large, quite elaborate leaves as do some conifers from the wet tropics. The familiar 'pine needle' (Fig. 7.6) may instead be a reflection of climate change, as the period which saw the greatest diversification of this group of plants was the relatively dry and cold Permian period. The continued success of conifers since then probably owes much to their ability to tol-

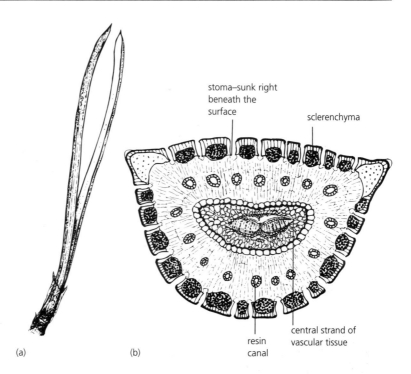

Fig. 7.6 A short shoot of the Scot's pine *Pinus sylvestris* (a) (×1.0) bearing two needle-like leaves. The tough, desiccation-resistant structure of the needles is shown in more detail in (b), a transverse section through a pine leaf, showing the sunken stomata.

(a)

(b)

erate the environmental extremes of the habitats where they still thrive.

Although not all conifers are 'evergreen' (*Larix*, the larch is one important and familiar example of a deciduous conifer), most do retain their foliage for long periods and have leaves capable of withstanding extremes of temperature and desiccation. This suits them admirably, as we shall see in Chapter 9, in areas prone to seasonal drought or cold. The record for longevity was until recently held by a bristlecone pine, and this species is one of the toughest organisms on the planet. One individual growing in the White Mountains of California is at least 4900 years old, and leaves may remain on the plant for 45 years. The wood of living and dead bristlecones provide an historic record stretching back over 8000 years. Because trees grow fastest and lay down thick rings of growth during warm wet years, analysis of tree rings can allow one to reconstruct the weather during their lifetime. The bristlecones are currently being used to analyse the climate changes which have occurred in the present interglacial period but the oldest plants on the planet may well be Wollemi pines—a relative of Monkey Puzzle trees until recently thought to be extinct. Some individuals of this Australian 'living fossil' are thought to be over 10 000 years old.

7.6.2 Reproduction

Conifer pollen

The robust vegetative features of the conifers are mirrored by their reproductive parts. The pollen of conifers is ideally suited to reproduction on dry land. Pollen grains are small, and some have air bladders (Fig. 7.7) which enhance their potential for long distance transport in the wind. So effective is this transport that pines have some of the greatest estimated distances of gene flow in the plant kingdom. This no doubt improves the efficiency of reproduction, but it may also be partly responsible for the low diversity of conifers; there are only 500 species. The high rates of gene flow may stop small populations from being reproductively isolated and forming new species.

Fertilization

Conifers do not have flagellate male gametes, so the last vestiges of reliance on a nuptial swimming pool

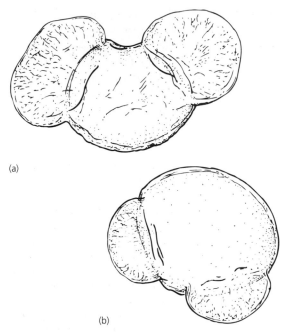

(a)

(b)

Fig. 7.7 Pollen grains of two species which help long-distance dispersal by possession of ear-like air bladders. (a) Scots pine *Pinus sylvestris*, (b) Norway spruce *Picea abies* (both ×1200).

for fertilization we witnessed in the cycads have been left behind. The pollen is carried right into the female cones by a combination of wind and then sticky fluid exuded by the ovule (a **pollen drop**). It then germinates to produce a tube capable of digesting its way through the megasporangial tissue (referred to as the **nucellus**) to the ovule. The mature grain has only four cells: two represent the remnants of somatic gametophyte cells and eventually disintegrate, one forms the pollen tube, and the other divides to form two sperm nuclei. When the pollen tube reaches an egg cell it discharges its contents and one of the two nuclei fuses with that of the egg. The other nucleus simply disintegrates, in complete contrast to that of the angiosperms, which has a very important role as we shall see later.

Seed dispersal

The female parts of most conifers are on the same trees as the males; unlike cycads, ginkgos and yews, they are **monoecious**. The ovules are tucked away inside cones, many of which seal up immediately

after pollination, protecting the upcoming generation still further. Indeed, some cones stay sealed until just the right conditions occur for seed germination. In the closed-cone pines it takes the heat of a fire to open the cones. The natural occurrence of a forest fire is therefore no disaster for these species, particularly as the ash formed by burnt parent plants provides rich mineral pickings for the germinating seeds. The cones of some other conifers open when mature to allow winged seeds to blow away on the wind, while some are prized open by birds or animals whose subsequent travels effect long-distance transport. Few conifers advertise for animal helpers, but the bright-red fleshy coverings of *Taxus* seeds (see Plate 6, facing p. 110) and the juicy berries of *Juniperus* provide rewards for dispersal agents, just like the fleshy fruits of some angiosperms.

7.7 THE ANGIOSPERMS (ANTHOPHYTA)

The angiosperms or flowering plants first emerged between 150 and 100 million years ago and have, in the relatively brief period since their arrival, come to dominate life on dry land. As we shall see over the last few chapters of this book, they have diversified into a dazzling array of over 230 000 different species, and have colonized almost all terrestrial and many aquatic habitats. Since much of the remainder of this book will concentrate on these plants, all we will consider here are the features that distinguish them from, and give them an evolutionary advantage over, the groups we have mentioned so far.

It is difficult to identify single characters that differentiate the angiosperms from all the other seed plants. In particular, there is a small group of rather unusual plants, the gnetales, that resemble angiosperms to a greater or lesser extent. We will meet several members of this group in the next few chapters. However, most angiosperms uniquely share a whole suite of characters that give them a competitive advantage over other plants in both their vegetative growth and reproduction.

7.7.1 Vegetative structures

Angiosperms share major novelties in both of their

major transport systems which undoubtedly improve their efficiency.

Modifications of the phloem

The phloem shows two major modifications from the condition in the conifers. Each phloem cell is accompanied by a living **companion cell** which is derived from the same mother cell rather than from the parenchyma as is the case with the **albuminous cells** of gymnosperms. The walls of each sieve-tube element of most angiosperms also contain a unique proteinaceous substance, known as **P-protein**, which may help seal the sieve-plate pores if the phloem is damaged.

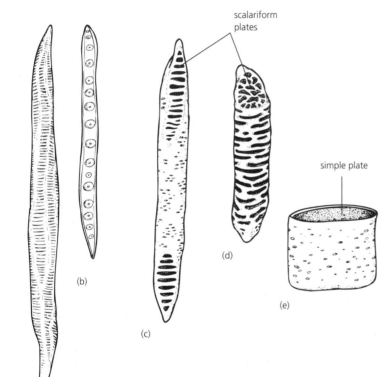

(a) (b) (c) (d) (e)

scalariform plates

simple plate

Fig. 7.8 Differentiation of xylem elements in angiosperms. Angiosperm wood contains long narrow **tracheids** and **fibres** (a, b) which have a mechanical role. Also present are the wider, shorter **vessels** (c–e) which are the main water-conducting elements. Vessels are joined end to end to form long tubes, either with semiopen **scalariform plate** ends (c, d) or completely open **simple plate** ends (e). The result is wood which is more efficient at transporting water but more prone to embolisms (all diagrams ×100). (f) Scanning electron micrograph of angiosperm wood, showing the wide but thin-walled vessels within the narrow but thicker-walled tracheids (×70).

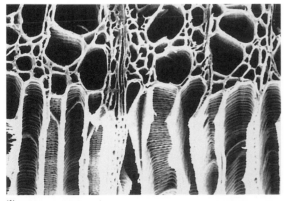

(f)

Differentiated xylem

The xylem of angiosperms shows great modifications. As we have seen, the water-conducting elements within the dense secondary xylem of conifers is made up only of **tracheids**. Angiosperm xylem is differentiated into very different cell types which have particular functions. Like conifers it contains long thin tracheids and **fibres**, under 30 μm in diameter (Fig. 7.8a, b), which have a mechanical role. However, it also contains much wider cylindrical elements called **vessel elements** (Fig. 7.8c–e), which can be up to

500 µm in diameter and which are stacked into continuous columns. The vessels are useless mechanically but because they are so wide, they have far less resistance to the flow of water. Angiosperm xylem is therefore much more efficient at transporting water and nutrient salts than that of most conifers. The only drawback of vessels, particularly long wide ones, is that especially in cold or dry conditions air bubbles or **embolisms** are likely to form, blocking water movement. Like other advances such as heterospory, vessels are thought to have arisen several times in evolution. They can be found in several of the most successful non-flowering plants we have already considered; *Pteridium*, *Selaginella* and *Equisetum*.

7.7.2 Reproduction

The advances made by angiosperms in their methods of reproduction have probably played an even more important role in their success. The most obvious feature is the possession of showy flowers, and the great development by angiosperms of insect pollination. However, by no means all angiosperms have the large colourful modified leaves of the most familiar species; it is the fine detail of the reproductive parts that has been critical.

Enclosure of the seed

First, the female parts, the ovules, are not just perched on top of sporophylls as in the conifers and the other gymnosperms ('naked seed' plants) but enclosed within an extra layer (Fig. 7.9). As we suspect for the integuments of the first seed plants, this layer is probably part of what was once a leaf blade, and is called a **carpel**. The area where the structure seals over is modified into a specialized surface (**stigma**) for the reception of pollen grains.

This enclosure was a critical stage in diversification of the angiosperms, because it opened the way for the evolution of chemical recognition systems between pollen grains and the stigma. These can act to prevent dissimilar pollen from germinating, and so can reduce the chances of hybridization and promote more rapid speciation. The other advantage is that as the seed undergoes development after fertilization, the carpel, and sometimes other maternal structures, can develop into **fruit** which protect it and may help its dispersal.

Reduction of the female gametophyte

Enclosure and protection of the female parts of angiosperms ('seed-in-a-vessel' plants) was accompanied by further reduction and simplification of the gametophytes. In the female parts of angiosperms there are no archegonia, and the single functional megaspore undergoes just three cell divisions (Fig. 7.10), producing a total of eight nuclei. The female gametophyte is therefore tiny and may be formed extremely rapidly, in a matter of days rather than weeks or months as in conifers. The eight nuclei in the gametophyte are arranged in two groups of four, at opposite ends of the megagametophyte (Fig. 7.10g). One nucleus from each group migrates into the centre of the eight-nucleate cell (Fig. 7.10h) and these two are described as the **polar nuclei**. The three nuclei nearest the micropyle form an egg cell and two accompanying cells. The three nuclei at the other end of the gametophyte also form cell walls, leaving the two polar nuclei in the middle to form a binucleate **central cell** (Fig. 7.10h).

Double fertilization

Whether the pollen arrives on the snout of a bat, the back of a bee, or from a passing breeze, the fertilization of angiosperms is different in one very important respect from that of gymnosperms. Pollen tubes enter ovules through the micropyle (Fig. 7.9), carrying two sperm nuclei, as before, but *both* nuclei now have important roles. One nucleus enters the egg cell, the other enters the central cell where it fuses with both the polar nuclei. This **double fertilization** sets the angiosperms apart from the gymnosperms. The fertilized central cell develops into a triploid **endosperm** (Fig. 7.10i), which is a tissue that will provide nutrients for the developing embryo. The advantage of this arrangement is that the seeds of angiosperms need be provisioned only if fertilized, and can continue to be supplied later in their development. In contrast, the embryos of conifers, like those of the ferns, are dependent upon material already laid down in the gametophyte for nutrition,

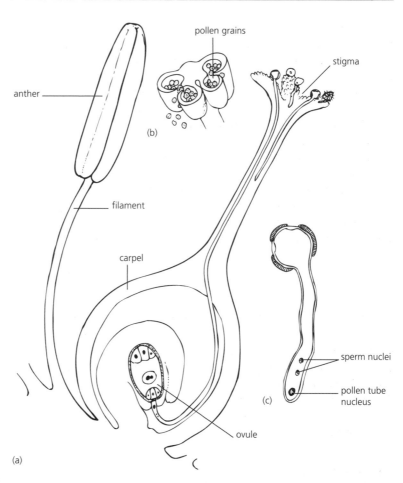

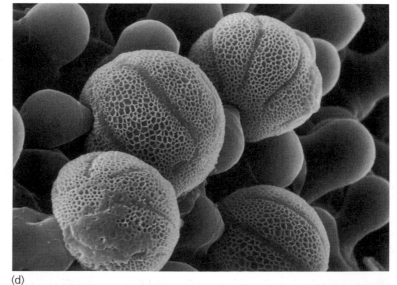

Fig. 7.9 Sections through the male and female parts of a typical angiosperm flower to show their characteristic features, and demonstrate the processes of pollination and fertilization. (a) Longitudinal section through the (female) carpel and (male) stamen. The tiny ovule is held at the bottom of the carpel. Pollen grains are released from the anther (see b) and land on the stigma to pollinate the flower. To fertilize it the pollen must germinate and grow a pollen tube which reaches right to the tip of the ovule. The pollen nuclei then travel down the tube to the ovule. There is **double fertilization**. One nucleus fuses with the egg cell to produce the embryo, another with the central cell to produce the endosperm. (c) Sections through pollen after germination. In the pollen tube can be seen the large tube nucleus and the small sperm nuclei. (d) Primula pollen sitting on a stigmatic surface (×2000).

121

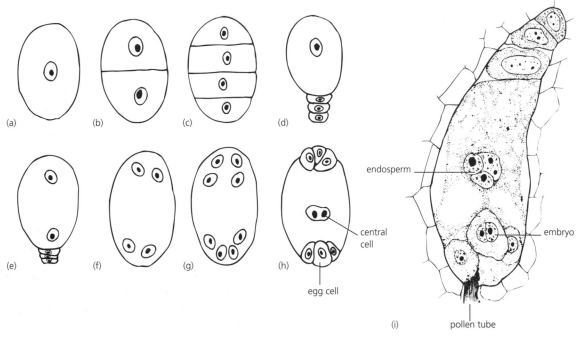

Fig. 7.10 Development and final morphology of the female gametophyte of an angiosperm. Idealized pattern of development (a–h) involves a complex set of divisions and cell death which produces eight cells (g). Two cells then migrate to the centre where they fuse and form the diploid **central cell.** (i) Mature female gametophyte of a lily *Lilium* at double fertilization (×1000). The pollen tube has brought the sperm nuclei, one of which is fusing with the egg to form the diploid **embryo**, the other fusing with the central cell to produce the triploid **endosperm**.

so must be provisioned *before* fertilization. The endosperm gives the angiosperms a similar competitive advantage in reproduction as the placenta gives to mammals.

7.8 THE GROUPS OF ANGIOSPERMS

Within the angiosperms, there are a few archaic groups, but the majority of plants belong to one of two major derived groups. There are around 165 000 species of dicotyledonous plants (dicots), which include the familiar trees, shrubs and many broad-leaved herbs. There are 65 000 species of monocotyledonous plants (monocots) which are almost all herbs and which include the lilies, irises, orchids and grasses. It is thought that the monocots diverged from the dicots around 100 million years ago.

7.8.1 Reproductive Differences

Monocots differ from dicots in several ways (Table 7.1 and Fig. 7.11). There are major differences in their reproductive characters. The flower parts of monocots such as the petals and sepals tend to be arranged in groups of three rather than in fours or fives as is common in dicots. The pollen grains are also different, being typically **monocolpate**, having only a single furrow or pore, rather than **tricolpate**, with three furrows or pores as in dicots. They also have only one seed leaf, or **cotyledon**, hence their name, compared with the two seen in most dicots.

7.8.2 Vegetative differences

However, the major differences that affect the form of the mature plants are seen in the vegetative characters. Monocots have a parallel arrangement of the veins in their narrow leaves rather than the net-like

Table 7.1 The main differences between the two major groups of the angiosperms.

Characteristic	Dicots	Monocots
Flower parts	Usually in fours or fives	Usually in threes
Pollen	Tricolpate	Monocolpate
Cotyledons	Two	One
Leaf venation	Usually net-like	Usually parallel
Vascular bundle arrangement	In a ring	Scattered
True secondary growth	Common	Rare

arrangement seen in most dicots. They also have a complex arrangement of vascular bundles in their stems rather than the single ring in which they are arranged in the stems of dicots. This latter feature means that the vascular bundles cannot be joined by a single continuous ring of cambium, and therefore monocots cannot easily undergo secondary thickening. This means that few monocots are able to grow into trees.

7.8.3 Major monocot groups

Palms are the main group of monocots that have evolved into trees, but because they have to lay down their wood during primary growth, even young specimens must have a very broad trunk, and their initial growth is necessarily slow. Just like tree ferns and cycads, they are therefore likely to be outcompeted by gymnosperms and dicot trees in most situations. Another problem is that because the trunk cannot thicken later in life, no new xylem vessels are laid down, and palms are therefore vulnerable to frosts, which cause embolisms in the vessels. It is probably for this reason that palms tend to be restricted to warm climates. Palms, like other monocots, need to produce adventitious roots from their stems as they grow to supplement the inadequate water supply, and many species produce obvious 'prop' roots which also supplement their anchorage.

Herbaceous monocots have adapted successfully to four major habitats: water, epiphytic life on branches, the floor of deciduous forests, and seasonally dry or cold habitats. It is difficult to see what competitive edge the differences confer on mono-

cots. Success in water may be due partly to their ability to undergo rapid extension growth and also to the arrangement of vascular bundles and veins which will improve the resistance of the long organs to tensile forces (see Chapter 11). As epiphytes, monocots such as orchids may have the advantage that they can rapidly produce adventitious roots which help absorb water (see Chapter 8).

In the other two habitats (see Chapter 9), success seems to be due to the ability of monocots to produce storage organs such as **bulbs**, which are made from expanded leaf bases, and **corms**, which are swollen stem bases. At the end of each growing season the cells rapidly divide to produce the next year's shoot which is packed into the storage organ. As conditions improve at the start of the next season these cells merely expand to produce a new shoot which is immediately ready to photosynthesize and reproduce. This gives these monocots a competitive advantage over other, slower-developing herbs. The arrangement also has the advantage of protecting the delicate **basal meristem** away from damage in the storage organ. This has been an important factor in the success of the grasses which in terms of biomass are the most successful of all the monocots.

7.9 THE EVOLUTION OF ANGIOSPERM FLOWERS

The evolution of the animal-pollinated flowers of angiosperms is a fascinating case of coevolution between totally unrelated groups of organisms. Animal pollination is not restricted to angiosperms.

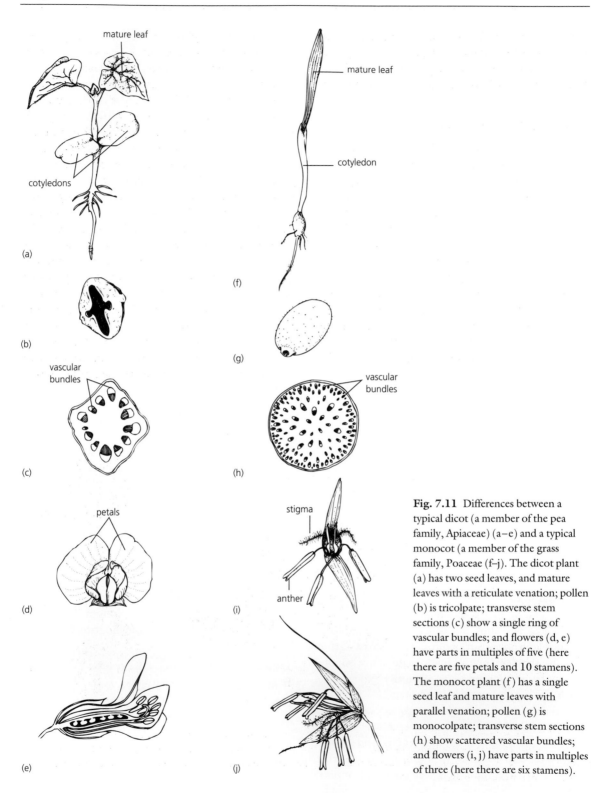

Fig. 7.11 Differences between a typical dicot (a member of the pea family, Apiaceae) (a–e) and a typical monocot (a member of the grass family, Poaceae (f–j). The dicot plant (a) has two seed leaves, and mature leaves with a reticulate venation; pollen (b) is tricolpate; transverse stem sections (c) show a single ring of vascular bundles; and flowers (d, e) have parts in multiples of five (here there are five petals and 10 stamens). The monocot plant (f) has a single seed leaf and mature leaves with parallel venation; pollen (g) is monocolpate; transverse stem sections (h) show scattered vascular bundles; and flowers (i, j) have parts in multiples of three (here there are six stamens).

Some modern cycads are insect pollinated as may have been some of the extinct gymnosperm group, the bennetitales. However, angiosperms show by far the greatest range of adaptations to animal pollination.

7.9.1 Origins of the angiosperm flower

There is little doubt that angiosperm flowers arose from unspecialized reproductive shoots possessing spirally arranged leaves and differentiated carpels or stamens. Even the flowers of some modern plants such as the angiosperm tree *Magnolia* and members of the buttercup family Ranunculaceae (Fig. 7.12) have the carpels and stamens arranged spirally. It was only later that the two lower layers of the typical angiosperm flower, the petals (or **corolla**) and the sepals (or **calyx**), differentiated from the stamens and leaves, respectively. This ancestral pattern is shown even by modern plants; they will produce undifferentiated leaf stalks instead of flowers if the supergenes which control flower shape are removed from the genome.

The earliest angiosperm flowers were probably partly wind pollinated, insect pollination also occurring when beetles which ate the protein-rich pollen accidentally transferred some as they travelled from flower to flower. This is certainly what happens during pollination in modern-day cycads (see Fig. 7.4), some of which heat up and produce enticing odours which attract insects. Subsequent evolutionary trends in flowers have acted to improve the efficiency of pollen transfer: the petals differentiated from the leaves as signals that would attract insects; sugar-rich nectar was produced which would act as a secondary attractant, particularly to female flowers; and the development of hermaphrodite flowers would have ensured better transfer between the sexes, although special mechanisms of self-incompatibility would have then been necessary to prevent too much self-pollination. These trends would have produced the 'typical' flower of Fig. 7.12. However, there have been several different directions in which flowers have altered from this basic design which have involved coevolution with different sorts of insects.

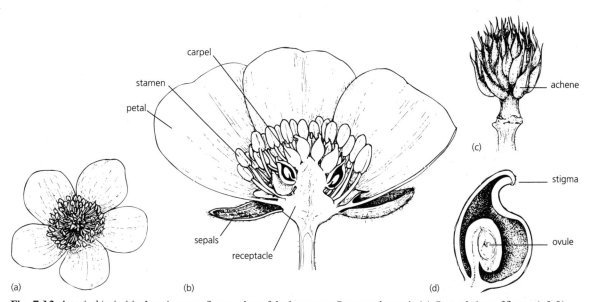

Fig. 7.12 A typical 'primitive' angiosperm flower, that of the buttercup *Ranunculus acris*. (a) General view of flower (×1.0). (b) Section through the complete flower, showing the four major whorls of flower parts: sepals, petals, stamens (male) and carpels (female). (c) Head of young nut-like **achenes**, showing their spiral arrangement. (d) Section through an achene to show the position of the ovule and stigma.

7.9.2 Trends in flower evolution

Increases in size

The first evolutionary trend has been an increase in size which helps attract more insects. One way of doing this has been to develop larger petals, and to form a globe-like flower in which insects such as beetles and flies can wander, eat pollen and be warmed by the sun. This is the path followed by the Magnoliaceae, Ranunculaceae and Rosaceae.

Grouping of flowers

Another trend has been the grouping of many small flowers into a large, flat **inflorescence** (Fig. 7.13i–l). The result of this trend is best seen in the **umbels** of members of the carrot family Apiaceae, such as cow parsley and, to its fullest extent, in the flower heads of the Asteraceae (see Plate 5a, facing p. 110) which are crowded onto a flat **capitulum**. The daisies in lawns are not single flowers with white petals but whole heads of yellow tubular disc florets surrounded by a ring of sterile white florets which have split along their length.

Changes in flower shape

An alternative trend has been the alteration of flower shape, which results in more efficient pollination by the animals which do visit. The increase in efficiency has been achieved by placing the reproductive elements closer together and altering the shape of the petals. Specialized pollinators can be attracted by producing greater nectar and pollen rewards, and producing petals of characteristic shape and colour.

Tubular flowers

One trend has been to fuse the petals into a narrow tube, at the entrance of which the tips of the stigma and stamens may be placed (see Plate 4c, d, facing p. 110, and Fig. 10.2e, f). Insects visiting such flowers are forced to touch both of them as they enter or

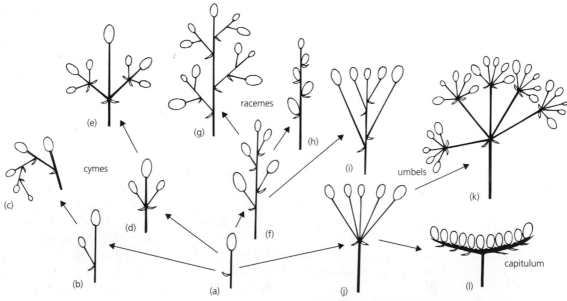

Fig. 7.13 The ways in which angiosperms arrange their flowers into **inflorescences** which help improve their attractiveness. In **cymes** (b–e) the terminal flowers open first and flowering proceeds from the top down. In **racemes** (f–i) the axial flowers open first and flowering proceeds from the bottom up. In **umbels** (j–k) flower stalks radiate from a single point, while in a **capitulum** (l) flowers emerge from a solid head. Holding flowers in a flat head (i–l) gives a platform for insects to walk between flowers.

stick in their tongue to suck nectar from its base. Subsequent coevolution between plant and insect will result in lengthening of the tube or of a basal spur. As a result fewer species of insect will visit the flower, and those that do must develop a longer tongue. Some tubular flowers of gentians have developed spurs over 30 cm long and become dependent on single species of insect. Such flowers are efficiently pollinated but at the cost of being wholly dependent for pollination on just a single species of moth or fly.

Bilaterally symmetrical flowers

Even more efficient transfer of pollen may be achieved by flowers that are bilaterally symmetrical and that insects therefore visit at a certain orientation (see Fig. 7.11d, e). The petals are fused into a tube with an obvious landing platform for the visitors. In such flowers the stigma and stamens can be arranged to touch the insect on its underneath, as in peas, or on its dorsal surface, as in labiates.

7.9.3 Coevolution of flowers and animals

Insect-pollinated flowers

As a result of the coevolution with insects, their main pollinator, flowers pollinated by different sorts of insects have evolved flowers of characteristic form and colour. Beetles prefer simple open flowers such as buttercups and roses, and umbels such as cow parsley. These flowers are usually white or yellow.

This is in sharp contrast to bee-pollinated flowers which are usually bilaterally symmetrical and coloured blue or purple. Good examples are lupins and thyme. Butterfly pollinated flowers, like lilacs, tend to be tubular and brightly coloured (Plate 4d, facing p. 110), while the night-flying moths visit tubular flowers with heavy scent such as honeysuckle.

Many flies are attracted by the small green or white flowers of species of *Euphorbia* (see Fig. 10.2c, d) and some of the less obvious orchids, which contain rewards of nectar and pollen. Other flies are attracted by the flowers of plants, such as the arum lilies and Rafflesias (Plates 5b and 7c, facing p. 110), which exploit their breeding behaviour. These plants pro-

duce unattractive brown or red flowers smelling of the rotten meat in which the flies usually lay their eggs.

Vertebrate-pollinated flowers

Flowers which are pollinated by other types of animal also have a characteristic form. Bird-pollinated flowers tend to be large and red in colour (Plate 4c, facing p. 110). Such flowers also tend to be tubular with projecting stigmas and stamens which touch the head of the bird while it feeds.

Flowers pollinated by mammals rely more on smell. The bat-pollinated flowers of rainforests (Plate 4a, b, facing p. 110) tend to be dull in colour but have strong fruit-like or musty scents, and are held on short strong shoots which can support the clumsy bats. Typical bat pollinated flowers include the economically important bananas, mangoes, kapok and sisal.

Orchids

There is no doubt that the plant group exhibiting the most spectacular adaptations for animal pollination are the orchids (Plate 4e, facing p. 110). All species have bilaterally symmetrical flowers with three petals and sepals but these show a great plasticity. Many species attract insects by providing nectar, and both fly pollinated orchids, with open green flowers, and bee-pollinated orchids, with tubular blue and purple flowers are common in temperate regions.

But orchids also use other techniques to attract insects. Some species supply oils or chemicals identical to the breeding pheromones of particular species of insects. Others, like members of the genus *Ophrys*, actually imitate females of particular insect species, causing the males to mistakenly mount them. The pollen, which in orchids is packaged into single **pollinia**, is stuck onto the visiting insect and has an excellent chance of being transferred to another flower of the same species when the insect attempts another conquest.

Many orchid species have developed relationships with just a single species of insect. This promotes efficient pollen transfer, which allows orchids to survive even if they are scattered widely in their habitat.

Development of such close relationships also promotes rapid speciation, resulting in the massive species diversity of the group. Unfortunately, though, because they are so reliant on one another, this leaves orchids and their pollinators particularly vulnerable to environmental change, and the fate of some orchids is heavily dependent on the survival of rainforests.

7.9.4 Wind-pollinated flowers

Not all angiosperms rely on animals for pollination. Many common species, like the grasses and many temperate trees, have reverted to wind-pollination. Grasses (see Fig. 7.11 and Plate 5c, facing p. 110) have tiny flowers with reduced petals, anthers exposed to the wind, and feathery stigmas to receive wind-blown pollen. Most of the trees have separate male and female flowers, the male flowers in many species being the familiar **catkins**, while the female flowers are reduced to simple feathery stigmas (Plate 5d, facing p. 110).

7.10 SEED DISPERSAL AND THE EVOLUTION OF FRUIT

7.10.1 Drawbacks of seeds

As dispersal devices, seeds have two major drawbacks. Because they contain such large energy stores to feed the growing seedling they provide attractive food sources for herbivores. Being much larger than spores they are also far less mobile and when released could simply fall straight down to the ground. Seed plants therefore have special adaptations to protect and disperse their seeds.

7.10.2 Arils and fruit

Certain gymnosperms have developed outgrowths of their seed coats, called **arils** to perform these tasks (see Plate 6c, d, facing p. 110 and Fig. 7.14). In contrast, angiosperms use tissue derived from the parent plant to produce **fruit**. Most fruit are developed from the carpels which surround the ovary, although some also contain material derived from the receptacle of the flower.

Because fruit develop from existing parts of the flower they can be produced more rapidly than the arils of gymnosperms and can form fully even before the seeds mature. This advantage has no doubt contributed still further to the success of angiosperms, and resulted in their development of a great diversity of fruit using a wide range of adaptations.

7.10.3 Adaptations for protection

The most basic role of fruit is to protect the seed from being eaten or from other mechanical damage. Among the most primitive types of fruit are the **achenes** of buttercups (see Fig. 7.11d), in which each seed is surrounded by a relatively unmodified carpel. The **nuts** of trees such as oaks and beeches are basically achenes in which the fruit wall has been strengthened by becoming thickened and extensively lignified. The shells of the strongest nuts, such as brazil nuts, have the densest sclerenchyma in the plant kingdom. There is only one problem with protecting the seeds mechanically in this way; it is hard for the seeds to germinate, and the nut must have some mechanism that allows the radicle to emerge. Most nuts have some sort of weakening which allows this, such as the ring between the two halves of walnut shells. Of course the danger with this is that seed-eaters can exploit these weaknesses.

7.10.5 Adaptations for wind dispersal

Almost all fruits have some protective value but most are more obviously adapted to aid dispersal. Wind is the most efficient dispersal mechanism for small seeds of up to a gram in weight.

Some, like those of poppies, merely use wind to push out the seeds through holes in the top of their **capsule**. Such seeds quickly fall to earth and do not disperse far. Seeds weighing under a milligram mostly use aerodynamic drag to slow their descent and allow them to apparently float away on the wind. The tiniest seeds, such as those of orchids and some fast-growing weeds, weigh only a few micrograms and need no particular adaptations to do this. Larger seeds, though, need to increase their surface area. Some, such as willows, attach hairs to the seeds or, as in the well known case of dandelions, make complex parachutes.

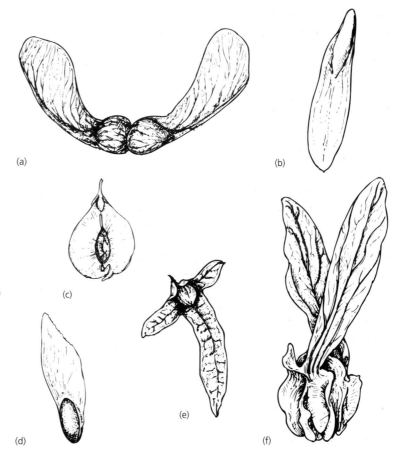

Fig. 7.14 Wind-dispersed seeds or **samaras** of trees. (a) Sycamore *Acer pseudoplanatus*, (b) ash *Fraxinus excelsior*, (c) wych elm *Ulmul glabra*, (d) Scots pine *Pinus sylvestris*, (e) hornbeam *Carpinus betulus*, (f) a rainforest dipterocarp tree *Dipterocarpus grandiflorus*. Note the convergence in form in wings formed by very different structures. The sycamore, ash, wych elm and dipterocarp samaras are formed from fruit, the hornbeam from a leaf, and the pine from the surface of a cone scale.

For seeds weighing above several milligrams, however, drag becomes inefficient at slowing their fall, so trees with larger seeds use lift-based mechanisms. The most common adaptation is to produce a fruit known as a **samara** which behaves like a helicopter, spinning as it falls (Fig. 7.14). This design has evolved several times, being seen in the well-known seeds of sycamores, maples and hornbeams. Pine trees have convergently evolved similar samaras (Fig. 7.14d) from outgrowths of the seed. Other arrangements with a varying number of blades have also evolved, and ash seeds (Fig. 7.14b) actually spin along their length as well as in a circle as they fall. All of these autogyrating samaras have a stable flight which suits them to windy conditions but because they fall vertically in calm conditions, none would be effective in the rainforest. To overcome this problem one rainforest vine from South-East Asia, *Alsomitra*

macrocarpa, has developed seeds that are in the form of a flying wing with a span of 15 cm. These are flung from the fruit by convection currents in the afternoon and glide slowly and gracefully to earth, hundreds of metres away.

7.10.6 Adaptations for explosive dispersal

Wind may be able to disperse small seeds from trees or plants growing in the open, but it is not useful for many forest herbs, which live in sheltered conditions. Many of these use explosive mechanisms to disperse their seeds. As the fruits of such plants dry, they tend to curl up, but because this is prevented by the structure of the fruit, mechanical tension builds up in the tissue. Eventually the fruit splits along prearranged lines of weakness, the fruit tissue quickly curls up, and the seeds are flung into the air. Good examples of exploding fruits are the capsules of bittercresses, violets, storksbills and Himalayan balsam. But the most widespread and well known are the exploding pods or **legumes** of the many members of the pea family, including those of the broom which explode with loud retorts on hot summer days.

7.10.7 Adaptations for animal dispersal

Hooked fruit

Despite the efficiency of other mechanisms, perhaps the most sophisticated forms of dispersal are those that use animals as the dispersal agent. The cheapest way of doing this is to attach the seed or fruit to a passing unwitting animal. Fruits with hooks have developed many times in evolution. The hooks may be derived from the style of the flower, as in herb bennet, the surface of the fruit, as in cleavers, or from other parts, as in agrimony. Whichever part is used, unwitting animal dispersal is very efficient and is used by around 10% of all angiosperms.

Edible fruit

However, by far the majority of angiosperms, and some gymnosperms, effectively bribe animals to disperse them (see Plates 6 and 7, facing p. 110). They coat their hard indigestible seeds with a soft, palatable coat which persuades animals to eat them. The seeds pass unscathed through the gut, before being deposited in a patch of nutrient-rich manure well away from the parent plant. Edible fruits have evolved many times and have a range of different structures.

Types of edible fruit. In **berries** such as tomatoes and grapes, the seeds are held within a fruit which has a soft inner wall, so the seeds require strong seed coats to allow them to survive. In **drupes** such as peaches, olives and cherries, on the other hand, each carpel holds only a single seed which is protected by the hard, stony inner layer of the fruit. Coconuts are another good example of such drupes, although we seldom see the fibrous outer layer of the fruit which allows it to float. **Pomes**, such as apples and pears, are fleshy fruits of the rose family in which the fleshy outer coat is derived from the receptacle. Only the core is derived from the carpel. Many fruits develop from more than one carpel. In strawberries (Plate 6a, facing p. 110), for example, the seeds are simple achenes on the surface of a fleshy receptacle. The red fruit of the yew tree *Taxus baccata* (Plate 6c, d, facing p. 110), is not a fruit at all, but a fleshy aril. It does not completely surround the seed, which is protected instead by being poisonous.

Colours of fruit. Over millions of years of coevolution with animals, all sorts of fruits have evolved basically similar mechanisms of ensuring dispersal. Most change colour to make themselves more conspicuous to their dispersers. Fruits which are dependent on birds for dispersal tend to be red, just like bird-pollinated flowers, but do not have much scent. This also explains the red colour of yew berries and of the seeds of many species of cycads which are dispersed by hornbills. Some plants make use of this attraction and colour their fruit red to mimic fruit (see Plate 6b, facing p. 110) thereby obtaining dispersal at no net cost.

In contrast to the conspicuous bird-dispersed fruit, many other fruits such as apples and pears tend to be inconspicuous and green in colour. Instead of birds, they attract small mammals to effect dispersal by emitting strong aromatic smells.

Ripening. The process of ripening also ensures that animals only eat fruits when the seeds are fully developed. Unripe fruits tend to be an inconspicuous green colour, lack scent, may have a disagreeable acidic taste and hard flesh and skin. As the seeds reach maturity, all of these characteristics are altered simultaneously under the influence of the hormone ethylene; the flesh is softened, the taste becomes sweeter and the change in colour or smell signal to animals that the fruit is ripe. In bananas much of the change occurs in the skin, which becomes softer, easier to peel off and changes from green to yellow in colour. Tannins, which make up 17% of the immature fruit, are also removed as it ripens.

Reducing the cost of dispersal. Of course the down side to bribing animals is the high cost to the plant of producing large fruit. Some plants seek to minimize their costs by bribing small animals which require less food. Many plants of deciduous forests, such as acacias, primroses, and violets are adapted to ant dispersal by attaching special food bodies called **elaiosomes** to the seed or fruit. The ants take the seed back to their nests to feed their young. The seeds themselves are unharmed and may germinate in the nest, obtaining both protection and nutrients in the bargain.

7.11 POINTS FOR DISCUSSION

1 Are there any *disadvantages* of producing seeds rather than spores?
2 Why do you think insect pollination is so rare in gymnosperms?
3 Mammals give birth to live young. Why do you think most angiosperms release dormant seeds rather than waiting for them to germinate?

4 Fruit bats are common in the Old World tropics. Can you think of any bat-dispersed fruit?

FURTHER READING

Bell, A.D. (1991) *Plant Form: an Illustrated Guide to Flowering Plant Morphology.* Oxford University Press, Oxford.

Cotter, F. & Sheffield, E. (1999) The Maidenhair Tree (*Ginkgo biloba*). *Biological Sciences Review* **12** (2), 13–15.

Proctor, M.C.F., Yeo, P. & Lack, A. (1996) *The Natural History of Pollination.* Harper Collins, London.

Rudall, P. (1987) *Anatomy of Flowering Plants.* Edward Arnold, London.

Stewart, W.N. & Rothwell, G.W. (1993) *Palaeobotany and the Evolution of Plants*, 2nd edn. Cambridge University Press, Cambridge.

Part 3

Part 2

Life in the wet tropics

8.1 INTRODUCTION: PROBLEMS OF LIFE IN THE WET TROPICS

Many areas of the tropics have a climate that is ideal for plant growth all the year round. Bright morning sunshine evaporates water from vegetation, driving convection currents which form clouds and produce rain in the afternoon and evening. The result is an equable climate which is warm, wet and humid with little wind. Such a climate would seem to be ideal for the growth of almost all the types of land plants we have examined in the last few chapters.

In these regions it is therefore not hard for plants to survive; their problem is that they must compete, with all the other plants that could survive, for limited resources such as light and nutrients. Consequently there will strong selection pressures on plants to grow as rapidly and as tall as possible to obtain light and shade out competitors. One might expect just a few species of the tallest possible trees, such as the conifers and angiosperms we examined in the last chapter, to win out. Trees with tall trunks and dense crowns ought to outcompete and shade out all other plants to produce tall, dense, species-poor **tropical rainforests**.

8.2 TROPICAL RAINFORESTS

The wet tropics are indeed dominated by trees just as one would predict. However, there are several surprises. For a start the vast majority of these trees are angiosperms, and there are very few conifers. But more intriguingly, rather than being dominated by just a few species of trees, rainforests show remarkable diversity. Rainforests not only contain a large number of species of angiosperm trees—one survey in the Amazon found over 500 species of large trees in a single hectare of forest—but also a wide range of plants with growth forms as diverse as **climbers**, **epiphytes** and short-lived **herbs**. Furthermore there are quite a large number of spore-producing plants and many bryophytes.

This chapter describes how the evolutionary novelties possessed by the angiosperms have allowed them to dominate the rainforests but also why some other plants also do well. Finally, it examines the reasons which have been put forward for why the rainforest flora is so stunning and so diverse.

8.3 VEGETATION OF THE PRIMARY RAINFOREST

8.3.1 Rainforest trees

The advantages of angiosperms

Angiosperm trees dominate tropical rainforests because their vegetative and reproductive systems both confer a decisive competitive advantage over conifers in the calm climate of the wet tropics. Because they possess wider, more efficient water-conducting vessels, angiosperm trees do not need such thick trunks for water transport as conifers. They can therefore grow tall much faster and invest relatively more material into productive photosynthetic machinery.

These trees are also pollinated efficiently by their animal partners which thrive in the calm conditions. For conifers, in contrast, the calm conditions would greatly hamper their wind pollination, and they would have to divert more energy into reproductive structures. Angiosperm trees are consequently by far the most common trees in rainforests. Conifers, such as the broad-leaved *Agathis*, are restricted to windier upland regions where, because all trees need to have broad trunks for mechanical stability and wind pollination is more efficient, conifers are less at a disadvantage.

Morphology of rainforest trees

In the ideal windless conditions of the rainforest, trees can grow up to 80 m tall. Most rainforest trees tend to look remarkably similar. They have narrow trunks with thin bark, and they are made of dense, darkly stained wood which contains few, large vessels. The trees branch to produce a crown only near the top, and their leaves are remarkably uniform, being thick and oval-shaped with extended **drip tips** (Fig. 8.1). The drip tips help the leaves to shed water as a stream of tiny droplets, which may prevent algae from growing on them, and reduce soil erosion around their roots.

Rainforest trees also tend to have characteristic root systems. The majority of their roots are superficial laterals which absorb nutrients from the thin humus layer, but sinker roots descend from the

laterals deep into the soil and anchor the trees like tent pegs. To prevent the lateral roots breaking under the strain this sets up, they are strengthened by the plate-like root buttresses which are such a feature of the rainforest.

Almost all rainforest trees are pollinated and dispersed by animals and have conspicuous flowers and fruit. Most flowers are held in the canopy, but in some trees, the flower stalks grow directly out of the trunk, a condition known as **cauliflory**. This allows them to produce larger flowers and fruit which may be pollinated by bats and dispersed by terrestrial animals.

Diversity of rainforest trees

Instead of a single 'optimal' species of tree, competition for the bright light has resulted in the evolution of many species which grow to different heights and which together produce a complex canopy (Fig. 8.2). There are three main sorts of trees: tall **emergent** trees, **canopy** trees and shorter **subcanopy** trees.

The leaves of the emergent trees are ideally placed for photosynthesis, being exposed to full sunlight and to the wind. But as a result the leaves lose large amounts of water by transpiration and, because they are so high, the trees can only draw enough water up the trunk to supply a small open canopy of narrow leaves. Consequently the emergents allow plenty of light through to other trees below.

The shorter canopy species are slightly shaded by the emergents and more sheltered from the wind. This means that they can support larger numbers of leaves and so can intercept a greater proportion of the remaining light. However, enough light can still penetrate the canopy to allow a few shorter and more slender subcanopy species to survive. Although many of these trees are also angiosperms, many forests, particularly in upland areas, have an understorey of tree ferns.

Understorey trees produce umbrella-like layers of leaves which intercept almost all of the remaining light. The three layers together therefore produce an efficient light-harvesting system, reducing light levels on the rainforest floor to only around 1% of its value above the canopy. In these dark conditions few plants can grow and the forest floor is a bare place

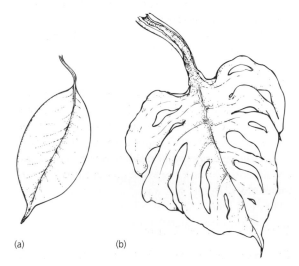

(a)   (b)

Fig. 8.1 Leaves of rainforest plants, showing their extended **drip tips**. (a) Leaf of a fig tree, *Ficus* sp. (b) Leaf of a climber, the cheese plant *Monstera* (both ×0.1).

Fig. 8.2 A diagrammatic cross-section through a tropical rainforest, showing the diversity of plants and the different layers of the canopy. The top layer (a) consists of the foliage of the **emergent** trees, with their open crowns. Note the buttresses helping to support the taller trees. Most light is absorbed by the crowns of the **canopy** tree layer (b) and the shorter **subcanopy** trees (c) which less often have buttresses. Below these layers the little remaining light supports layers containing only a few tree ferns and herbs (d) and seedling trees (e). Note the presence also of climbing woody **lianas** and **epiphytes** on the trunks and branches of the taller trees. On the right-hand side of the illustration an emergent tree has fallen over, leaving a canopy gap which is being colonized by herbs like gingers, by fast-growing lianas and rattans and by **pioneer** trees.

dotted only with small seedling trees which stay permanently stunted (unless a mature tree falls and light can reach them) and a very few shade-tolerant ferns and club mosses, including members of the genus *Selaginella*. Despite this, however, large plants of two other life forms do manage to thrive in the forest.

8.3.2 Climbing plants

The second most important group of plants in the rainforest are the woody climbers, or **lianas**, which reach the light by effectively parasitizing the structural investment of the trees. Climbers may twine up

narrow trunks and branches, grip on with tendrils, or attach themselves to tree trunks using adventitious roots, but whichever technique they use they will not require a wide stem. They will have lower maintenance costs and so will be able to survive and grow even near the forest floor, until they eventually reach the canopy.

Selection pressures on climbers

The main problem with climbing up trees is that it becomes difficult to supply the leaves with water through the long narrow stem. This means that only plants which have developed xylem vessels can

Fig. 8.3 The climbing gnetale *Gnetum* (a) (×0.1), a close relative of the angiosperms, possesses xylem vessels and angiosperm-like leaves. However, it lacks the typical angiosperm petalled flowers, instead having separate female branches (b) (×0.5), shown here with mature ovules, and male branches (c) (×2), shown here with developing microsporangia.

become climbers and even these must produce wider vessels than trees. Most woody climbers are therefore angiosperms. The exceptions include the 50 or so lianas of the genus *Gnetum* (Fig. 8.3). These plants are members of the Gnetales, a small group of plants with close affinity to the angiosperms, which seem to have developed some convergent adaptations to them. We will meet other members in Chapters 9 and 10. Other climbers are ferns, such as *Lygodium*. Both *Gnetum* and *Lygodium* have evolved vessels independently. There are no climbing conifers at all.

Evolution of climbers

The climbing habit has evolved many times independently in the angiosperms. Lianas are spread around more than 20 families but they all show some degree of convergence, not only in the xylem vessels but in the overall structure of the stem. Most lianas develop some form of **anomalous secondary thickening** in which several woody segments are dissected by soft parenchyma tissue. Together with the twisting of their stems this gives them rope-like flexibility which allows them to survive even if their supports move or collapse beneath them. Economically important species of lianas include the rattans from which cane furniture is made. These are actually climbing palms which scramble up vegetation using a mass of hooks. There are also climbing bamboos.

Strangling figs

An intriguing group of lianescent plants are strangling plants which includes the strangling figs, of the genus *Ficus* (see Plate 7a, facing p. 110), and members of the genera *Clusia* and *Schifflera*. They start out life as epiphytes before developing into climbers and finally free-standing plants. The young figs germinate in the barks of their host and produce climbing stems and numbers of aerial roots which grow down to the forest floor. Over a period of 50–100 years the twining stems and roots gradually get thicker and thicker, coalescing and strangling the host tree, which eventually dies. The fig, however, carries on growing, the aerial roots anchoring it firmly in the ground and forming a sort of lattice-work trunk. As the trunk of the host tree rots away this trunk gradually takes over the support of its developing crown and the fig becomes a free-standing tree.

8.3.3 Epiphytes

A third group of plants, the **epiphytes**, also effectively parasitize the structural investment of trees, but this time by living on their trunks and lower branches.

Selection pressures on epiphytes

In some ways this habitat is ideal for plant growth. Epiphytes are secure from terrestrial herbivores and are protected by the canopies in which they live from extremes of light and heat. The main disadvantages, which result from the lack of soil, are the lack of nutrients and the danger of desiccation. For this reason this niche is dominated, as elsewhere in the world, by the poikilohydric lichens and bryophytes discussed in Chapter 5. In the rainforest, however, the dangers of desiccation are reduced by the high humidity and rainfall, and therefore the niche can be colonized by other plants as well. The most successful groups are desiccation-tolerant ferns like those we encountered in Chapter 6 and certain angiosperms which have evolved special adaptations to increase water and nutrient uptake and reduce water losses.

Adaptations of epiphytes

Many of the endohydric plants show physiological adaptation to reduce water loss just like the desert plants we will examine in Chapter 10. Seventy per cent of the angiosperm species and many ferns, such as *Pyrrossia*, photosynthesize using crassulacean acid metabolism (CAM) which allows them to close their stomata during the day (see also Chapter 10). So similar, indeed, are the adaptive forces to conserve water in epiphytes and desert plants that the Amazonian rainforest contains many epiphytic cacti!

Many epiphytes also show morphological adaptation to capture water and nutrients. The stag's horn

fern *Platycerium* and the bird's nest fern *Asplenium nidus* (Fig. 8.4a, b) both create a basket of leaf bases which helps to trap water, dead leaves and other debris. This trend is much further developed in many bromeliads (Fig. 8.4c) whose overlapping leaves form a water-tight tank. Water and nutrients which are captured by this tank are absorbed using specialized hairs or **trichomes** on the leaf surface. This mechanism is even better developed in the air plants of the genus *Tillandsia* which absorb rainwater and dew over their entire surface using trichomes (Fig. 8.5) which open up when the surface is wet to allow water in, and close when it dries to prevent water loss. The most extreme development is seen in Spanish moss, *Tillandsia usneoides*, (Fig. 8.4f) which shows striking morphological convergence with the lichen *Usnea* (see Chapter 5) after which it was named.

A further adaptation seen in many epiphytes is to form mutualistic relationships with insects. Many epiphytic ferns in particular develop swollen chambered stem bases (Fig. 8.6) in which ants can make their nests. In return for shelter the plants obtain nutrients from their excreta and decaying bodies. Some bryophytes have the best of both worlds. The two-lobed 'leaves' of the *Pleurozia* shown in Fig. 8.7 bear one highly modified lobe. This sac-like structure is invaginated, with an entrance tunnel which has 'doors' opening into the sac. This is a water-holding sac with a one-way valve, but also serves as a trap for insects—once they are in, like the water, they stay there!

Epiphytic orchids

But the most successful of all the groups of epiphytes are the orchids, possibly because the family possesses so many preadaptations to the epiphytic way of life. Orchids produce tiny seeds which can easily colonize branches, and have an especially close mutualistic relationship with fungi which helps them obtain nutrients. Their exceptionally efficient insect pollination (see Chapter 7) also allows even scattered individuals to reproduce successfully. Epiphytic orchids have also developed two other specializations for life aloft. Many possess thickened stem tubers (Fig. 8.4d) that store water. Some have also reduced their

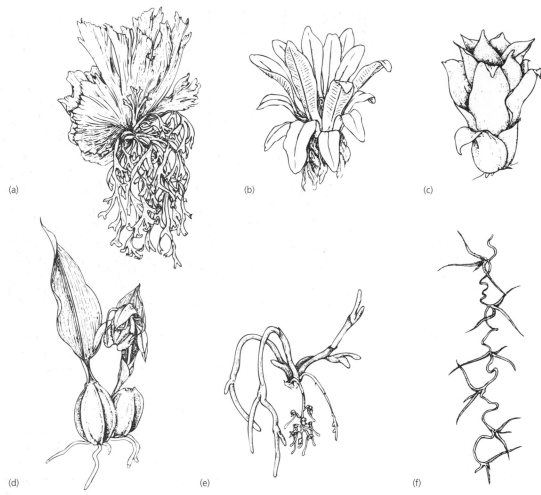

Fig. 8.4 Epiphytes of tropical rainforests. (a) The stag's horn fern *Platycerium* (×0.1) has two sorts of frond: broad brown fronds (above) which form a basket which traps falling water and leaves; and the lower green photosynthetic fronds. (b) The bird's nest fern *Asplenium nidus* (×0.05) whose fronds form a more regular basket. (c) A tank plant *Vriesia* of the family Bromeliaceae (×0.15), an angiosperm whose leaves form a waterproof water-storing tank. In the orchid *Coelogyne* (d) (×0.2) water is instead stored in the swollen stem base. Reduction in epiphytes is seen in (e) *Microcoelia microglossa* (×0.2), a leafless orchid which photosynthesizes though its green roots, and in (f) the rootless air plant *Tillandsia usneoides* (×0.3), which absorbs water through mobile scales on its leaves (see also Fig. 8.5).

leaves (Fig. 8.4e) and carry out their photosynthesis in their large roots. These roots are covered by a multi-layered epidermis called the **velamen**, a structure also possessed by epiphytic arums. The roots take up water readily when wet, allowing the cells of the velamen to conduct water inwards. However, in desiccating conditions the velamen dries, disconnecting the water link between the central root and the surface, and so reducing water loss.

8.3.4 Rainforest parasites

A final group of plants steal not just the support that would enable them to photosynthesize, but the products of photosynthesis itself. Parasites like the arum lily *Rafflesia* (see Plate 7c, facing p. 110), which grows in the rainforests of South-East Asia, obtains its carbohydrates, water and nutrients directly from the roots of vines. To do this it produces a

Fig. 8.5 Scanning electron micrograph of the surface of *Tillandsia usneoides* (×100), showing the thick covering of mobile scales through which it takes up water.

haustorium which plugs into the xylem and phloem stream of its host. With a direct supply of water and nutrients assured, parasites require no leaves, roots or stems to survive. Consequently all *Rafflesia* produces apart from its haustorium is its huge flower, the largest single flower in the world.

8.4 VEGETATION IN GAPS

Together, the competing trees, climbers and epiphytes produce a deep complex canopy which harvests the light so efficiently that all other plants should be excluded. But however long lived it is, every plant has a finite lifespan. Each tree will eventually die, collapse or fall over, and these catastrophes will leave an open gap in the canopy. It is in these gaps that a completely different range of shorter-lived plants may, for a time, thrive.

8.4.1 The microenvironment of gaps

The microenvironment of gaps is very different from that under the intact canopy; there is much more light and during the day it is much hotter. In these conditions, faster-growing light-loving plants will have an advantage since they can outgrow and smother competing vegetation. Therefore there will be selection for plants which maximize growth and minimize investment in structural materials.

8.4.2 Vegetation of new gaps

There are two ways in which plants can reduce such investment. The first is to reduce the amount of woody material they lay down and instead to support themselves using turgor pressure. The second is to parasitize the efforts of other plants by climbing up them. As a result gaps are initially colonized by non-woody plants (or **herbs** as they are known), particularly climbers.

Rainforest herbs

Herbaceousness is restricted among vascular plants to the spore-producing forms, particularly ferns, and to the angiosperms. This is because these groups are both capable of reproducing extremely rapidly before they get outcompeted in their site. Ferns and their allies quickly produce tiny spores; and the seeds of angiosperms can develop in a matter of days. As we have seen, of course, some ferns and angiosperms also developed xylem vessels which prompted the evolution of herbaceous climbers. In contrast, it is not surprising that, given their slow method of

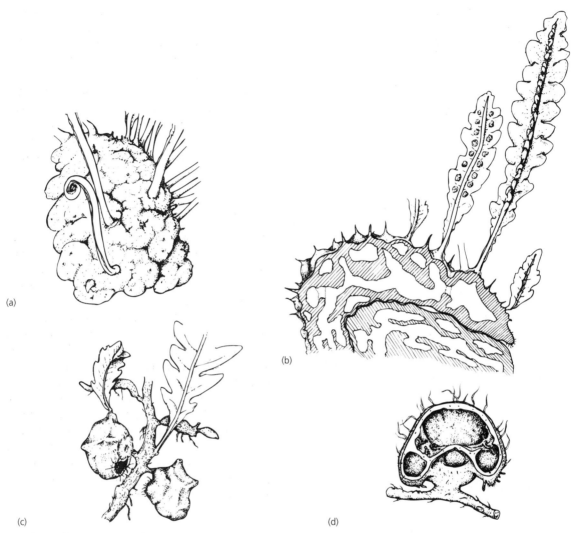

Fig. 8.6 Ant ferns of tropical rainforests. Members of the genus *Lecanopteris* harbour ants in a swollen rhizome. (a) Plan view of *L. carnosa* (×0.3). (b) Diagrammatic section of *L. spinosa* showing the hollow chambers which are inhabited by the ants. Members of the genus *Solanopteris* produce swollen tubors to hold the ants. (c) Plan view of *S. bifromis* (×0.5). (d) Section through tubor of *S. brunei*.

reproduction and lack of vessels, no herbaceous or climbing conifers have evolved. As we shall see in Chapter 9, herbaceous angiosperms are even more common in areas with seasonal climates, but gaps in the rainforest (and the roadside verges which mimic such gaps!) are still dominated by a range of herbs, including gingers, grasses, climbing ferns, horsetails, and scrambling club mosses of the genus *Lycopodiella*.

The mass of vegetation in new gaps can also be created artificially by humans when roads are built or clearings made around a village. The hopeless tangle of vegetation that springs up is called a jungle. Although dissimilar to the vegetation in the primary rainforest, western colonialists' familiarity with these areas gave rise to the myth that rainforest is impenetrable.

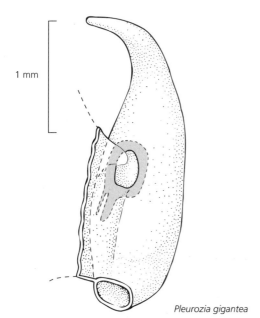

1 mm

Pleurozia gigantea

Fig. 8.7 Most 'leafy' liverworts have small bi-lobed leaf-like structures, but this one has the under-lobe developed into a complex inflated sac-like structure with an invaginated entrance tunnel. The tunnel (shown grey in the diagram here) ends in a pair of neat inward-facing castanet-like doors which act as a one-way valve into the sac, so that water (or small animals) can enter but not leave.

8.4.3 Succession in gaps

Growth of pioneer trees

Herbs dominate gaps initially but they cannot grow particularly tall because of the high costs of maintaining turgor pressure in the live cells which help support them. Consequently they become overgrown by a number of species of **pioneer trees** which, although they produce wood like other trees, have a range of adaptations to speed up their growth and outcompete the other plants. The wood that trees such as the South-East Asian *Macaranga* and Amazonian balsa *Ochroma* lay down is very light and requires little carbon investment. The trunks of South American *Cecropia* trees are even lighter as they are hollow, with thin walls and strengthening bulkheads, just like the structures of oil tankers! Pioneer trees develop just a few large branches from the top of their trunks, and these support a small

umbrella-like canopy of huge leaves. This arrangement has two advantages. Competing climbers have no lower branches to support their growth and so cannot grow up them, and other herbs are shaded out by the shallow canopy. Some pioneer trees also harbour ants in their hollow twigs (see Plate 8a, facing p. 110) in a fascinating mutualistic relationship. They provide the ants with shelter and nectar; in return the ants kill and eat herbivores which might eat the tree's leaves and bite through the growing tip of vines which might strangle them.

Re-establishment of climax trees

After 10–20 years, gaps therefore become dominated by pioneer trees. These trees continue to grow and reproduce, shedding their seeds to colonize new gaps. The partly shaded conditions below the pioneer trees are, however, perfect for the growth of the larger canopy **climax trees** which eventually outgrow the shorter-lived pioneers and, after 50–100 years, replace them. This process, which is known as **secondary succession**, results in the eventual reestablishment of the original forest structure. A tropical rainforest is really a dynamic mosaic of patches of forest at different stages of this succession, allowing the rainforest to contain yet more types of plant. Even so, the profligate diversity of the tropical rainforest remains baffling to tropical ecologists and many reasons have been put forward as to why they contain so many species of trees.

8.5 POSSIBLE CAUSES OF HIGH RAINFOREST DIVERSITY

8.5.1 Perfect growing conditions

The most obvious explanation for the high diversity of rainforests is that in such perfect growing conditions many species can survive. But, as we have seen, good conditions merely result in greater competition between plants, which one might expect to be won by only a very few species. Good growing conditions should perhaps promote the sort of low diversity seen in the temperate forests we will examine in the next chapter.

8.5.2 The great age of rainforests

A second possibility is that more species have evolved in rainforests simply because they have had more time to do so. Rainforests have indeed existed for tens of millions of years, and, being equatorial, were far less influenced by the recent ice ages than were the plant communities of higher latitudes. Whereas, in the northern temperate zone, vegetation types migrated thousands of miles southwards during each ice age, rainforests were merely reduced to small refugia, from which they could readily expand again as conditions improved. Even today these refugia retain particularly high diversity.

8.5.3 Interspecific competition

Another possible explanation for the high diversity involves competition between the plants. It suggests that local differences in the environment create different niches to which particular species can become adapted. Trees may be adapted to invade larger or smaller gaps, or may thrive in soils of slightly different composition. Unfortunately there is little evidence to back up this idea; these explanations ought to apply just as much to the species-poor temperate forests.

8.5.4 Herbivory and pathogens

Perhaps the explanation lies not in competition between trees but with their interactions with other organisms. The large number of herbivores, particularly insects, which survive in the equable climate may prevent any one tree from becoming too common. A common tree would present insects with a constant supply of readily available food. The numbers of herbivores specializing on it would consequently build up and frequently defoliate it.

Common trees would also be subjected to frequent attacks by pathogenic fungi and viruses. Herbivory and pathogens can impose strong selection pressures on plants to produce novel defensive chemicals. It is perhaps not surprising therefore that in this environment such pressures have driven plants to produce a particularly wide range of defence compounds. Many are used by the indigenous peoples of rainforests for medicinal purposes. This has recently encouraged drug companies to set up large programmes of research to try and identify compounds which can be used to fight cancer and other diseases.

8.5.5 Animal pollination

A final possibility is that the animal pollination which is common in the rainforests may have promoted speciation. Because most pollinating insects do not travel very far to feed, two populations of trees can easily become reproductively isolated from each other. Once separated geographically, genetic drift can cause the populations to diverge. Adaptation of separated populations to being pollinated by different species of insects can also rapidly cause the flower morphology to change, so causing speciation.

8.5.6 Conclusion

More research clearly needs to be carried out to determine the true cause of diversity in rainforest trees. It may be due to a combination of factors. However, whatever the cause, the high diversity of plants in tropical rainforests is probably in turn responsible for increasing the diversity of animals and so making them so fascinating to makers of wildlife documentaries.

8.6 THE EFFECT OF HUMANS ON RAINFORESTS

The devastating effect humans have had on tropical rainforests this century has been well documented. Rainforests continue to be destroyed at an alarming rate and soon there may be little primary rainforest left outside nature reserves. Much anger has been directed at the developing nations whose rainforests are suffering. However, the reasons why they have not protected their rainforests as well as they might are understandable. Many of the reasons are actually similar to those that long ago led humans to destroy almost all the primary forests of Europe and North America.

8.6.1 Causes of rainforest destruction

Expanding native populations cut down forests to provide firewood and to open up more land to grow their own food. Trees may also be cut down by farmers to make way for commercial plantations, or by cattle ranchers to make way for pasture. Finally, governments or logging companies may cut down trees to sell the timber. Cutting down trees to make way for agriculture generally destroys the rainforest ecosystem but, surprisingly, logging need not do so.

The felling and harvesting of trees produces areas of disturbed vegetation which are effectively huge gaps. These areas will be colonized by herbs and pioneer trees just like smaller gaps in primary forest, to create a species-poor tangle of vegetation. Gradually, though, the pioneer trees will shade out the herbs and in turn be replaced by climax trees which have grown in their shade; eventually a **secondary forest** will be created which is similar to the original primary forest, if usually somewhat shorter and less diverse.

8.6.2 Sustainable use of rainforests

In areas with low pressure on land use, rainforests may therefore be progressively harvested in a sustainable way, just like the broad-leaved woods of temperate areas. Much conservation effort is increasingly being channelled into developing methods of reducing logging damage and speeding up the recovery of rainforests. Our best hope for the future of rainforests is probably that we can learn to use them sustainably, producing forests that are both economically viable and diverse.

Wood is not the only resource that makes rainforests valuable. They can also be exploited, as we have seen, for medicinal chemicals and fruit. But perhaps the fastest-growing resource is human curiosity. Well-managed ecotourism centres are springing up in rainforests throughout the tropics, bringing in valuable foreign currency and educating local people about their natural heritage.

8.7 POINTS FOR DISCUSSION

1 If all angiosperms died out, which plants do you think would take over the rainforests? How do you think they would evolve in the future?
2 How do you think recovery of rainforests after logging might be speeded up?
3 Why don't temperate trees produce such big buttresses as those in rainforests?

FURTHER READING

Archibold, O.W. (1995) *Ecology of World Vegetation.* Kluwer Academic Publishers, Amsterdam.

Luttge, U. (1989) *Vascular Plants as Epiphytes.* Springer-Verlag, Berlin.

Putz, F.E. & Mooney, H.A. (1991) *The Biology of Vines.* Cambridge University Press, Cambridge.

Richards, P.W. (1996) *The Tropical Rain Forest.* Cambridge University Press, Cambridge.

Whitmore, T.C. (1990) *An Introduction to Tropical Rainforests.* Oxford University Press, Oxford.

Life in seasonal climates

9.1 INTRODUCTION

9.1.1 Latitude and seasonality

The equatorial belt, where the rainforests flourish, is the one area on earth which has an equable climate. In all other parts of the world plant growth is restricted for at least a short period each year by seasonal changes. Climate changes vary fairly predictably with latitude.

In the outer tropics, plant growth is restricted during the short **dry season** which occurs when the earth is tilted away from the sun. The dry season becomes longer and more predictable as latitude increases, and so plant growth in many subtropical regions can occur only during the short summer **monsoon**. At the still higher latitudes of the desert belts, around 25–35 degrees, the dry season lasts all year round!

Above 35 degrees latitude the growing conditions improve. The climate becomes predictably damp in the winter, particularly on the west-facing coasts of the **Mediterranean** regions of south-west Europe, California and Western Australia. Above around 45 degrees, in the temperate zones, the climate becomes damp throughout the year and plant growth is only restricted by the cold and relative darkness of winter. At higher latitudes, however, particularly in the subpolar regions above 60 degrees the growing season becomes shorter again, as winter lengthens. Finally, in the polar regions, above 70 degrees, winter lasts all year round and conditions are always inimical to plant growth.

The climates of all these regions also differ in two other respects from that of the central tropics. They are much more windy and they all have much higher temperature differences between day and night.

9.1.2 Latitude and plant distribution

Clearly, latitude will influence the sorts of plants that can survive, but the effect of latitude on plant distribution is not a simple one. Instead, areas with growing seasons of *similar length* tend to have similar vegetation. Partly this is because drought and cold both affect plants in the same way; they put them under water stress, either because there is no water in the soil for the roots to take up or because the water is frozen solid. In addition, of course, the low light levels and cold of winter will also directly reduce growth. As we shall see, plants growing in seasonal areas have developed a range of adaptations that prevent water loss by seasonally reducing growth. This chapter examines how seasonality affects plant morphology and investigates its effects on the phylogenetic composition and ecology of the vegetation.

9.2 DISTRIBUTION AND ADAPTATIONS OF TREES

9.2.1 Areas where trees can survive

Three main areas have long growing seasons: the outer tropics with their long wet season; the Mediterranean regions, with their long wet winters; and the outer temperate zones, with their long wet summers. In all of these areas, trees are able to survive (Fig. 9.1), so even outside the wet tropics large areas of the world are still covered by forests. However, the trees and other plants they contain are rather different from

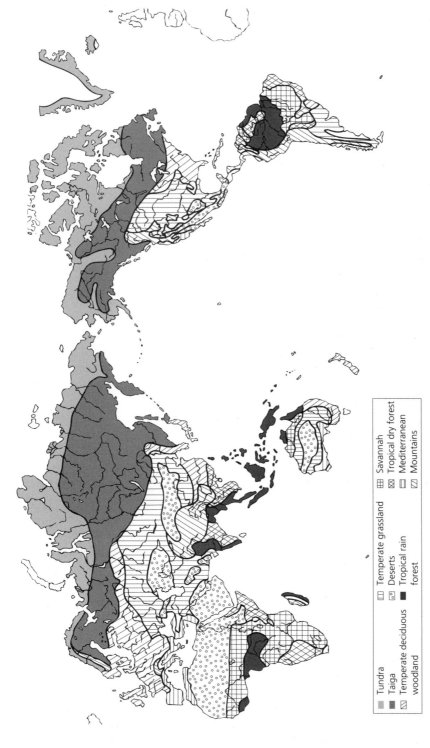

Fig. 9.1 Map of the world showing the distribution of the main vegetation types. **Tropical rainforests** tend to dominate the central tropics, and tropical dry forest areas further away from the equator. Trees also dominate many temperate regions; warmer, wetter areas develop temperate deciduous woodland, colder areas develop **taiga**, or conifer forest and seasonally dry areas develop Mediterranean woodland. Trees are absent in the dry areas around the desert belts at latitudes of 15–40 degrees. Here there are tropical grasslands called **savannahs**, deserts and **temperate grasslands**, sometimes called **prairies**. Polar regions are dominated by **tundra**.

Legend:
- Tundra
- Taiga
- Temperate deciduous woodland
- Temperate grassland
- Deserts
- Tropical rain forest
- Savannah
- Tropical dry forest
- Mediterranean
- Mountains

those of the tropical rainforests. Rather than being dominated by angiosperms, seasonal forests may contain angiosperm or coniferous trees, or sometimes both. The two groups show an interesting mix of convergent adaptations and non-convergent adaptations which help them survive the dry or cold season.

9.2.2 Convergent adaptations of angiosperms and conifers

Both angiosperms and conifers have developed several morphological features that help them resist drought, temperature and mechanical stresses. Both sorts of tree protect their delicate apical meristems by encasing them during the dry or cold season in hard drought-resistant scales. The scales form the outer covering of the **perennating buds** (Fig. 9.2) which are packed with soft preformed shoots. The buds are a clever adaptation which allows the trees to unfurl their new shoots extremely rapidly when good growing conditions return. All the cell division has been completed in the previous growing season, so that at bud-burst the preformed cells only have to expand.

There are two main advantages of this arrangement: not only can photosynthesis be resumed as soon as possible, but leaves can harden and lay down woody material sooner, minimizing the time when they are vulnerable to being eaten by herbivores.

Both angiosperms and conifers of seasonal environments also have thick coverings of bark, which reduces water loss and insulates the cambium from rapid temperature fluctuations. Finally, trees in both groups also have thicker trunks and branches than rainforest trees, which helps them withstand the mechanical forces imposed by strong winds.

9.2.3 Reversion in angiosperms

In many ways, angiosperm trees have reverted to morphology which is similar to that of conifers. Their wood contains a large number of narrow xylem vessels which are less prone to being blocked by embolisms in dry or cold weather than the large vessels of rainforest trees. Many angiosperm trees have reverted to wind-pollination (see Plate 5d, facing p. 110), which is just as, or even more, effective than

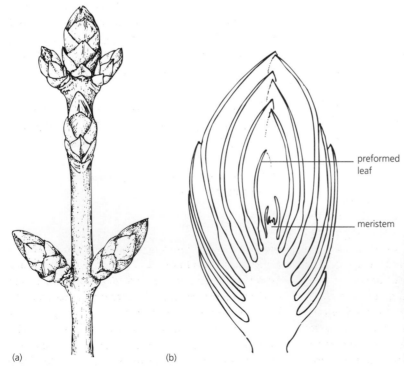

(a)   (b)

Fig. 9.2 Perennating buds of the sycamore *Acer pseudoplanatus*, a deciduous tree from north-west Europe. (a) Plan view, showing the drought-resistant scales. (b) Diagrammatic longitudinal section through a bud, showing the preformed leaves and meristem, protected beneath the scales.

preformed leaf

meristem

insect pollination in cold, windy conditions. Many trees also have wind-dispersal of their seeds (see Fig. 7.14, p. 129). In both pollination and seed dispersal the degree of convergence between angiosperm trees and conifers can be striking; trees like alders produce female cones which are very similar to those of conifers, and the samaras of pines (Fig. 7.14d) look and behave just like those of sycamores (Fig. 7.14a).

The consequence of these adaptations is that, unlike in the rainforest, angiosperm trees no longer enjoy a clear competitive advantage over conifers. Both require thick trunks, have narrow water-conducting elements, and are wind pollinated.

9.2.4 Non-convergent adaptations

Trees have developed two alternative sets of adaptations that enable them to conserve water during the dry season, just like the ferns we examined in Chapter 6. The first, which is used by most conifers, is to produce drought-resistant foliage. Pine needles are a good example of weatherproof but highly efficient photosynthetic structures (see Fig. 7.6, p. 117). They have a high surface to volume ratio, therefore absorb light well, and have excellent gas-exchange properties, but are double-wrapped with a thick cuticle and a layer of specially thickened cells which protect the all-important mesophyll. Such structures are metabolically expensive to make, so many species hold onto their leaves for long periods. Bristlecone pine needles remain photosynthetically active for up to 45 years. Similar adaptations are seen in the small evergreen leaves of some Mediterranean angiosperms.

The second set of adaptations, which is more commonly employed by angiosperms, is to shed the foliage at the end of the growing season. **Deciduous** trees possess a specially weakened layer of cells at the base of the petiole called the **abscission layer** (Fig. 9.3). At the end of the growing season this layer is digested away, allowing the petiole to snap off. The leaf scars are then covered by a layer of corky tissue. Losses of energy and nutrients are reduced by transporting organic substances and chlorophyll from the leaves, back into the body of the tree. This gives many trees their characteristic autumnal coloration, which is caused by the predominance of the yellow

and red carotenoid pigments and anthocyanin which are jettisoned with the leaves. Because a deciduous tree must produce a completely new canopy each year, rapid growth of new leaves is even more important than in evergreens. Many deciduous trees consequently have particularly large buds containing many preformed leaves.

9.2.5 Advantages and disadvantages of evergreen and deciduous trees

Both of the water-conservation strategies described above have advantages and disadvantages. Evergreen trees do not have to invest time and resources at the start of each year to produce new leaves and can photosynthesize whenever conditions are good enough. However, the productivity of each leaf tends to be lower because its adaptations to reduce water loss also reduce the rate at which carbon dioxide can enter for photosynthesis. Each leaf also requires more resources to build. The leaves of deciduous trees are generally thinner and more flexible giving them three main advantages. They are cheaper to construct, capable of higher rates of photosynthesis, and can also roll up in the wind to reduce drag. This last mechanism is particularly effective in lobed leaves, like those of sycamores, and in pinnate leaves, like those of ash and locust trees, and helps reduce wind damage to both the leaves and the tree.

9.3 VEGETATION OF SEASONAL FORESTS

9.3.1 Distribution of evergreen and deciduous forests

The effect of climate

Neither the deciduous nor the evergreen strategy is better in all conditions and the type of tree which dominates a particular area will depend on the details of the climate (see Fig. 9.1). Deciduous trees tend to have an advantage in areas with very long or very good growing seasons where the high productivity of their leaves outweighs the annual losses due to leaf fall. Deciduous trees therefore tend to dominate

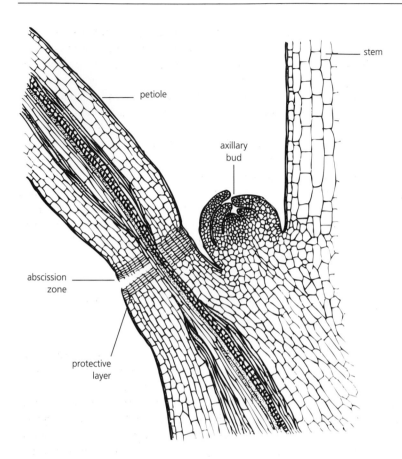

Fig. 9.3 Abscission in a deciduous leaf. Transverse section through the base of a petiole in *Acer pseudoplatanus* (×7). Cells in the abscission zone are orientated to form a line of weakness. Cell walls are digested away at the end of the growing season, allowing the petiole to detach, and a protective corky layer is formed at the end of the stump.

most subtropical **monsoon forests**, where most rain falls in the hot summers. They also do well in many temperate areas such as western Europe and the eastern United States which have long wet summers, forming **temperate deciduous forests**, dominated by oaks, beeches, limes and ashes.

Evergreens will have an advantage in areas with a shorter or poorer growing season. These include areas with a Mediterranean climate where the growing season is the cool winter: the Mediterranean itself, where forests of pine and evergreen oak are found; and California where the great redwoods grow. Evergreen conifers also dominate huge areas of the cooler northern temperate regions where the summers are short. Much of Canada, Scandinavia and Russia are covered by the conifer or **boreal forests** which are also known as **taiga**.

The effect of altitude

Altitude also has a major effect on the distribution of trees, because of its effect on climate. Even at latitudes where deciduous trees are normally favoured, evergreens can outcompete them in mountain areas. These include not only the alpine regions of temperate latitudes, which have short summers because of their elevation, but also tropical mountains. These areas are not truly seasonal but show massive diurnal variations in climate. Temperatures can rise above 30°C during the day but can fall below freezing point at night, so a montane tree can encounter in a single day as much variation as a temperate tree would encounter in a year. No plant can afford to drop its leaves every single day so the best strategy is to be evergreen!

Other environmental effects

Other environmental factors as well as climate also affect the pattern of tree distribution. Evergreens tend to be favoured, for example, on poorer soils. The advantages and disadvantages are not always clear cut, and so most habitats include both evergreen and deciduous trees. Such mixed forests include the native Caledonian forest of Scotland, which is a mixed forest of Scots pine *Pinus sylvestris* and birch *Betula pendula*, while maples are common understorey trees in coastal Canada.

There also seem to be some exceptions to the general rule about climate, the most notable being the distribution of the deciduous conifer, the larch *Larix siberica*, which covers huge areas of north-eastern Siberia where one might expect only evergreens to thrive. Perhaps the winter cold and drought here is too severe for even conifer leaves to survive.

9.3.2 Tree size

Despite the adaptations they have to make to withstand desiccation and higher winds, trees in seasonal forests can grow even larger than those in tropical rainforests. The tallest trees in the world are temperate species. The conifer *Sequioa sempervirens*, for instance, grows up to 110 m tall in its native habitat of coastal California and south-west Oregon. Many angiosperm trees, such as the eucalyptus trees of Tasmania, grow to comparable heights. Generally, trees growing in sheltered areas and with a good water supply grow taller than those in windy drought-prone environments.

9.3.3 The scarcity of lianas and epiphytes

Although trees grow well in seasonal areas, lianas and epiphytes find the conditions much more difficult to withstand. Because woody climbers require very wide vessels to transport adequate water up their narrow trunks, they are particularly vulnerable to drought and especially cold weather; the water column in the vessels is readily broken. Lianas therefore become increasingly rare at higher latitudes and at higher altitudes. One of the few species that does thrive is ivy, *Hedera helix*, which climbs with the use

of tiny adventitious roots. These not only grip the tree's bark to provide support but also suck up water close to where it is needed, reducing the need for excessively wide vessels.

Epiphytic angiosperms also become much rarer in seasonal conditions, because they are particularly vulnerable to desiccation during dry spells. Away from the rainforests, the epiphytic niche is dominated by plants that can survive severe desiccation: bryophytes, lichens and some ferns. One of the few angiosperms that does survive in these conditions is mistletoe (see Fig. 9.6c below) which is actually a **hemiparasite**. Its roots tap into the xylem stream of it host, so obtaining a constant supply of water and nutrient salts and avoiding desiccation.

9.3.4 Ground vegetation in seasonal forests

Deciduous forests

Compared with rainforests, seasonal forests, especially deciduous forests, have much more ground vegetation. This is because beneath deciduous trees, conditions are ideal for plant growth at least during one short period each year—at the onset of the growing season before their canopies have developed. During this period the forest floor is well lit and is sheltered from extreme temperature fluctuations by the twigs and branches of the trees. However, in such a short time even herbaceous angiosperms cannot develop and reproduce from seed. Instead the forest floor of deciduous woodlands is dominated by long-lived **perennial** herbs, such as the forest ferns we introduced in Chapter 6 and many angiosperms.

The key to the survival of perennials in this habitat is their development of underground storage organs such as tap roots, corms or bulbs (Fig. 9.4), on which the perennating buds are held. The soil protects these organs from desiccation and extremes of temperature, as well as from predation by large herbivores. At the start of each growing season the plant mobilizes some of the energy stored in the organs, rapidly expanding their cells which have already been produced, and so opening their leaves. Perennial herbs can photosynthesize enough before the trees come into leaf not only to produce a flower stalk and to set seed but also to replenish the energy stores and

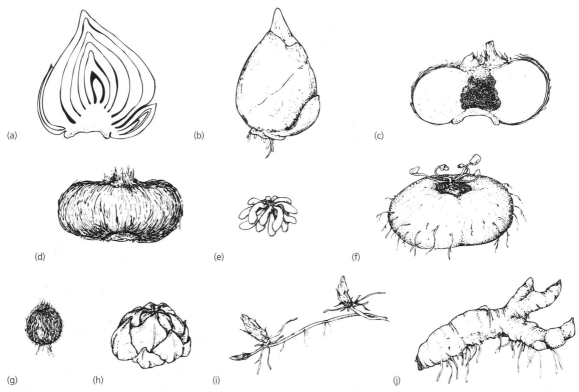

Fig. 9.4 Storage and vegetative reproductive organs of herbaceous perennial plants from seasonal habitats such as woodlands and Mediterranean areas. (a, b) Section and lateral view of tulip (*Tulipa*) bulbs (×0.5), formed from swollen leaf bases, (c, d) corms of *Gladiolus* formed from a swollen stem base, (e) tubers of buttercup *Ranunculus* (×1.0), (f) tuber of *Cyclamen* (×0.1). (g) Corm of *Crocus* (×0.6), (h) bulb of *Lilium* (×0.3), (i) rhizome of *Convallaria* (×0.5) and (j) rhizome of *Iris* (×0.5), formed from modified stems. Bulbs, corms and tubors store food well but dispersal is better carried out by rhizomes.

produce new perennating buds. The leaves can then die back, leaving only the dormant underground storage organ to survive until the next year.

So successful have been perennial angiosperms that the floor of temperate deciduous woods is a riot of colour. Some of the plants such as anemones and primroses are dicotyledonous. But the niche is dominated by monocots whose suite of adaptations for rapid growth (see Chapter 7) gives them a decisive advantage, allowing plants such as trilliums in North America and bluebells and wild garlic in northern Europe to carpet the ground.

Evergreen forests

In contrast to deciduous forests, the floor of evergreen coniferous woodlands is permanently shaded,

so the ground vegetation is sparse. Only a few shade-adapted plants such as bryophytes and ferns may survive, and therefore coniferous woods, and especially plantations, are much less attractive places in which to walk.

9.3.5 Parasitic plants

A final way of obtaining enough energy to grow and reproduce in the dark conditions beneath the canopy is to steal carbohydrates. Hemiparasitic plants such as broomrapes and orchids (Figs 9.5 and 9.6) are fairly common in seasonal forests, just as in tropical rainforests. They obtain their carbohydrates from the roots of trees using modified roots which tap into the phloem stream of their hosts. Because they do not need to photosynthesize, many do not produce

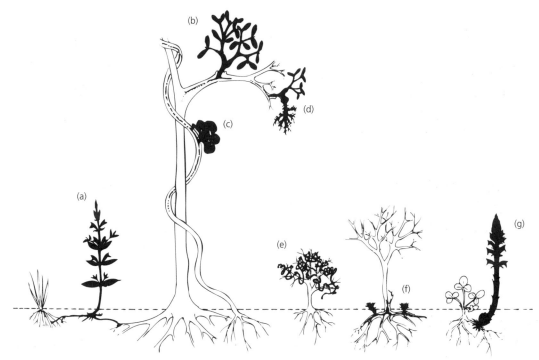

Fig. 9.5 Types of parasites. **Hemiparasites** (a, b). (a) A terrestrial hemiparasite (family Scrophulariaceae, (b) an epiphytic hemiparasite (family Viscaceae). **Holoparasites** (c–g). (c) *Rafflesia* and (f) *Cytinus* (both Rafflesiaceae), holoparasites that are mostly **endophytic**, growing within the host. The **epiphytic** holoparasites: (d) family Loranthaceae, (e) the rootless *Cuscata* and (g) the root parasite *Orobanche*.

leaves, and lack chlorophyll, and hence their flowering stems have a characteristic brown coloration. Some orchids (see Plate 7d, facing p. 110) are even **saprophytic**, obtaining energy from rotting tree roots via their mycorrhizas.

9.4 ECOLOGY OF SEASONAL FORESTS

9.4.1 Succession

Despite the differences in their flora, seasonal forests regenerate in a similar way to rainforests after disturbance. Gaps are filled by a secondary succession of short-lived herbs scrambling plants such as brambles, most of which are angiosperms, followed by shrubs, pioneer trees and finally by climax trees. However, the process is altered to some extent by the seasonal conditions.

Early succession

Gaps are initially colonized by extremely fast-growing **annual** herbs which put all their resources into setting seed before the end of the growing season. However, because they must re-establish from seed each year, annuals are outcompeted in subsequent years by plants that can survive between growing seasons. Many of these plants are herbs that survive by investing in a permanent root system which acts as an energy store. **Biennial** herbs such as foxgloves store the energy from the first year's growth in a tap root which powers the growth of the flower spike in the second year. Even these, though, are outcompeted by **perennial** herbs like willowherb, which produce new shoots each year from their underground perennating buds. Of course, in order to survive and spread, all these herbs must produce reproductive propagules before their foliage dies down each year. This seems to be a major reason why, just as in

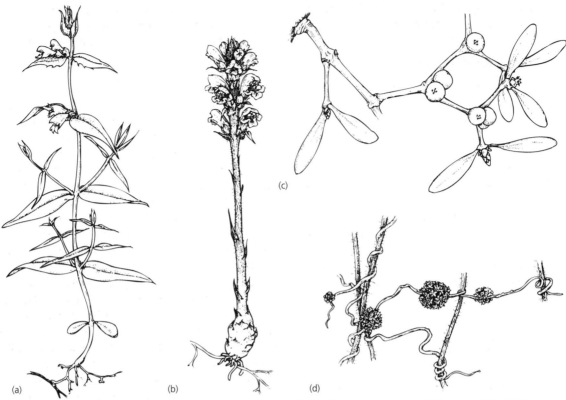

Fig. 9.6 Holoparasites and hemiparasites from temperate areas. The hemiparasite cow wheat *Melampyron* (a) (×0.5) is a root parasite obtaining water and nutrient salts from the xylem stream of trees and shrubs. The parasite *Orobanche* (b) (×0.4) absorbs all its nutrition from the roots of shrubs and herbs. The mistletoe *Viscum album* (c) (×0.7) is a hemiparasite on branches. Both (a) and (c) are green and are capable of photosynthesis. Dodder *Cuscuta europaea* (d) (×1.0) is a holoparasite on the stems of nettles. Neither (b) nor (d) has any leaves or photosynthetic apparatus.

tropical rainforests, there are no herbaceous conifers, and why ferns and angiosperms dominate early succession.

Late succession

Because woody plants maintain a permanent structure above ground, their shoots have a head start at the beginning of each growing season. For this reason, although they grow slowly at first, woody plants eventually shade out the temporary stems of the herbs. The first woody plants to dominate are **shrubs**, which hold a dense canopy of leaves on a number of rapidly branching stems, and which can therefore grow relatively rapidly. However, because

they do not have a trunk, their height is limited. They will therefore eventually start to be shaded out by **pioneer trees**.

The pioneer trees of the seasonal woods are quite different from their rainforest counterparts. They have much denser wood which may help them cope in the windier conditions. Perhaps because there are few woody climbers to compete with, they do not produce the single layer of large leaves seen in the rainforest pioneers. Instead, they typically have large numbers of small leaves which are arranged in a deep open canopy. This maximizes their growth rate, but since such a canopy casts little shade, climax trees, which have a flatter but more closed canopy, are able to carry on growing beneath them. Eventually the climax trees outgrow the pioneers and reassert the status quo, just as in the rainforest.

9.4.2 Diversity of seasonal forests

Seasonal forests have a much lower species diversity than tropical rainforests. Partly this is because the canopy is simpler. There are usually only two layers, the canopy and the subcanopy. As we have seen, there are also fewer climbers and epiphytes, although there are more plants on the forest floor.

But the main reason why seasonal forests are less diverse is that they tend to be dominated by only a very few species of trees. The reasons for this have not been adequately explained, but it may be related in part to the high incidence of wind pollination. This often results in high levels of gene flow between populations which helps prevent speciation.

9.5 THE CLIMATIC LIMITS OF TREE GROWTH

9.5.1 Problems with short growing seasons

As we have seen, three main areas of the world have short growing seasons: the subtropics with its short wet monsoons; the lower temperate regions, with their short wet winters; and the subpolar regions, with their short summers. In these areas, conditions are less favourable for the growth of trees. As the growing season becomes shorter and less productive it becomes more difficult for trees to produce and maintain their large trunk and branches. In drier, windier conditions trees also lose more water from their exposed leaves and find it harder to replace it with water from the drier soil. Fires will also kill their exposed meristems.

9.5.2 Tree morphology

In these more difficult conditions trees show reduced growth and more extreme adaptations to reduce water loss. The dominant trees tend to be species which in better areas would be pioneer trees. This is because their deep, open canopies use water more efficiently than do the monolayer canopies of climax

species. Their lower leaves are shaded from desiccating sunlight by the upper ones and so, despite photosynthesizing at an adequate rate, lose less water.

9.5.3 Replacement of trees

As conditions deteriorate still further, the trees increasingly have to grow longer root systems to obtain an adequate supply of water and nutrients. Consequently, they grow further apart, and gaps open up in the canopy. More light penetrates to the ground and can power the growth of other types of plants which require less energy and water to survive. It is for these reasons that forests give way to other types of vegetation in areas with short growing seasons, which, as we shall see, include huge areas of the globe (see Fig. 9.1).

In the subtropics, forests give way to **savannahs**; in the lower temperate regions, to **prairies** or **steppes**; and in subpolar regions to **tundra**. Particularly dry habitats in other regions may also end up treeless. In drier, rocky areas of the Mediterranean regions, for instance, tree cover breaks up to allow growth of scrubby **maquis** vegetation; and temperate areas with porous, nutrient poor sandy soils can support only **heathland** vegetation.

9.6 VEGETATION IN AREAS WITH SHORT GROWING SEASONS

9.6.1 Life history of successful plants

Despite their very different climates all the areas described above support remarkably similar vegetation. Instead of trees, the dominant plant forms are ones which in better areas would appear earlier even than pioneer trees in the succession: shrubs and perennial herbs, most of them angiosperms.

There are many reasons why these types of plant are better able to survive here than trees. Being shorter and having less biomass than trees, they will be less prone to water loss and will have lower maintenance costs. Perennials have the additional advantage that their underground buds are better protected from desiccation, the cold, herbivores and fires. Angiosperms dominate this niche, probably because they can flower and complete seed production within a single growing season.

9.6.2 The grasses

The most successful of all the plants in these regions are the grasses (family Poaceae), which are monocots, and mostly perennial. Their narrow, vertically orientated leaves are well suited to bright light. They photosynthesize rapidly, while avoiding direct exposure to the sun, and are efficiently cooled by the wind. Grasses can spread asexually using underground rhizomes. But the adaptations that have probably contributed most to their success relate to their defences against **herbivory**, a crucial aspect of life for such short vegetation.

The leaves of grasses have large numbers of sclerenchyma fibres running up their length. These toughen the leaves and make them less profitable to eat, but they also prevent herbivores from tearing them off. Herbivores have to adopt slower methods. Small herbivores such as insects and rabbits must carefully cut all the way across the leaf, while cattle adopt rougher tactics by gripping the leaves and pulling them upwards until they break. Even if the grass leaves do break this will not stop the grass from growing because extension growth, as in many monocots, occurs in **basal meristems** which are well protected deep in the foliage.

But the most sophisticated defence of grasses is their incorporation within their epidermal cells of silica bodies which wear down the teeth of herbivores when they chew them. Since their first appearance some 30 million years ago grasses have been involved in an evolutionary arms race with grazing mammals such as ungulates and rodents. The grasses have incorporated more and more silica, and developed more and more complex crystal shapes, while the ungulates have developed longer and more complex molar teeth. This is why the jaws of modern horses and cattle are so deep—the skull is virtually all teeth! The process has been carried on even further by rodents such as voles whose incisor teeth carry on growing throughout their life to replace the wear caused by constant gnawing.

With all these advantages grasses have come to dominate savannahs and prairies. Today, grasslands cover over a quarter of all the dry land. So common indeed are they that grasses have reverted to wind pollination (see Fig. 7.11, p. 124, and Plate 5c, facing p. 110). They produce flower stalks which project above the leaves into the wind and the inconspicuous green flowers open to allow the anthers to project. These release their pollen into the air and it is intercepted by the feathery stigmas of the female part of other flowers. Despite the lack of specificity of their pollination mechanisms, grasses are extremely diverse; the grass family Poaceae is second only to the orchids in numerical terms, with over 20 000 species. Some grasses have even managed to grow as tall as trees, despite their lack of secondary thickening. Bamboos, for instance, can grow to a height of up to 40 m thanks to the efficient tubular design of their stems and their habit of growing in dense stands where each stem is supported by its neighbours. The three most important food plants used by humans—wheat, maize and rice—are also grasses.

9.6.3 Perennial dicots

As well as grasses and their close relatives, the sedges (Cyperaceae), grasslands also contain numerous dicots, including members of the daisy family (Asteraceae) and peas (Fabaceae). The daisies no doubt benefit in these regions from their effective chemical defences (see Box 9.1), while the peas have an advantage in the extreme root competition because of the symbiotic nitrogen-fixing bacteria in their root nodules.

Most of the dicots show some convergence in form to the grasses, possessing narrow, vertically orientated leaves, while many are pollinated and dispersed by the wind. Similarly, like grasses many show morphological adaptations to C_4 **photosynthesis**, which increases their productivity in the hot climate. C_4 plants hold their chloroplasts in their **bundle sheath cells** which surround the veins, and transport carbon dioxide to them by a chemical shunting mechanism. This mechanism speeds up photosynthesis at high temperatures, because it keeps oxygen away from the photosynthetic reactions with which it would interfere. Water loss is also minimized because the stomata do not need to open so wide to allow in carbon dioxide.

BOX 9.1 CHEMICAL DEFENCES

Types of defences

One of the most surprising things in nature is the survival of plants despite attack by thousands of species of herbivores. After all, plants are unable to run away. Throughout their evolutionary history there has therefore been extremely strong selection pressure on plants to evolve defences. They use four main strategies to avoid being eaten: they can avoid herbivores, make themselves hard or unprofitable to eat, enlist help from other animals, or make themselves poisonous.

We have already seen many examples of the first three forms of defence: many perennial herbs, especially monocots, have hidden underground meristems; spines, hairs and silica crystals make leaves of many plants difficult to eat; and pioneer trees of rainforests harbour friendly ants to ward off herbivores. But the most widespread defences used by plants are chemical, and the evolution of novel poisons have been key events in the evolution of the angiosperms. Plants have developed two major forms of chemical defence: quantitative and qualitative defences.

Quantitative defences

Quantitative chemical defences are used by a wide range of plants, and act to make them difficult to digest. The major groups of chemicals which perform this function are **tannins** and **phenolic** compounds, both of which act by precipitating the digestive enzymes in the stomach. Large quantities are required, however, and therefore the mechanism is energetically expensive. The tannins which give flavour to tea leaves, for instance, are found in concentrations of up to 2%.

Many plants reduce the costs by producing these chemicals only if they have been attacked. Oak trees, for example, produce leaves with higher levels of tannins and phenolics if attacked by gypsy moths. These responses have recently been implicated in the notorious migrations of lemmings, which feed on Arctic sedges. When there are few lemmings around, sedges suffer little damage so do not produce defence compounds. However, when the lemming population rises, the damage to the sedges increases and it becomes profitable for them to produce the compounds. Lemmings that eat the tainted sedge cannot digest their food and begin to starve, causing them to migrate in massed hordes in a doomed attempt to find new pastures.

Qualitative defences

A metabolically cheaper defence is the production of small quantities of poison. Many ferns and gymnosperms produce highly toxic compounds. Many plants, bracken is an example, are cyanogenic (they produce cyanide when grazed). However, toxins have been most widely studied in the angio-sperms, in which the different families rely on different classes of compounds.

Members of the pea family (Fabaceae) use cyanogenic glycosides which produce hydrogen cyanide when the plant is bruised, while the figworts (Scrophulariaceae) produce cardiac glycosides such as the digitoxin found in foxgloves. The cabbage family (Brassicaceae) rely on glycosinilates or 'mustard oils' which give them their bitter taste and characteristic smell. The drawback is that brassicas lack symbiotic mycorrhizas, possibly because of the toxic effects of these compounds. As a result they tend to be short-lived plants restricted to ephemeral sites.

However, the most deadly of the compounds produced by angiosperms are the alkaloids, which are found in the borages (Boraginaceae) and daisy family (Asteraceae). Many alkaloids, like many other plant toxins, have useful pharmacological properties. Atropine, which is produced by deadly nightshade, is a useful nerve toxin, and other well-used alkaloids are morphine, codeine, caffeine and nicotine. It is likely that many such compounds remain undiscovered in the rainforest, or may be known only to the native people. Rainforests may be vital medicine cabinets to the world.

Detoxification and coevolution

The main problem with relying on toxins is that herbivores may evolve the ability to withstand or detoxify them. Cinnibar moths, for example, thrive on ragwort and actually sequester the senecionin that the ragwort produces and which can kill other insects and even cattle. They use it for their own defence, advertising this with warning black and yellow coloration. Monarch butterfly caterpillars do much the same thing with the cardiac glycoside toxin produced by their milkweed host plant. In fact most plant poisons can be detoxified by at least one species of insect, which is why so many species of insect have specific 'host plants' that they eat. Some, like cabbage white butterflies, even use the smell of the toxin to detect their host plant. The effectiveness

(*Continued on p. 158.*)

Box 9.1 contd.

of a particular compound will therefore decrease over evolutionary time.

So important is the coevolution of plants and herbivores that it is thought that they have been an important driving force in the emergence of the successive families of angiosperms. As each new group emerges with a novel poison it will diversify rapidly before herbivores evolve resistance. It is then challenged by new groups which have produced new ranges of compounds. One of the most currently successful angiosperm families, the Asteraceae, is probably merely the most recent with the most novel compounds.

9.6.4 Bulbed perennials

Because water or nutrients limit plant growth more than they do in forests, there is stronger selection pressure on plants to obtain and defend these resources. Consequently there is more competition between plant root systems, and the soil is held in a tangled web of roots. In drier areas there are particularly strong selection pressures for plants to finish their growth and reproduction as quickly as possible in the growing season, before the water runs out. Here, bulbed monocots, which as we have seen are well adapted to grow in deciduous forests, are well adapted for survival and can even outcompete grasses.

9.6.4 Hemiparasites

So great is the competition for water and nutrients that some plants even steal them from other plants. Many members of the figwort family Scrophulariaceae, which flourish in grasslands, are hemiparasites (Figs 9.5 and 9.6 and Plate 7b, facing p. 110), tapping into the xylem stream of grasses using specialized roots called haustoria.

9.6.5 Hedgehog plants

Defence is a particular problem for the shrubs which live in highly seasonal climates because their meristems are exposed above ground even when they are not growing. These plants must therefore put particularly high investment into this activity. Many of these plants are highly poisonous, but most also have highly effective mechanical defences. The **hedgehog plants** of the Mediterranean have developed such defences to their ultimate extent, the foliage being held within what is effectively a cushion of spines.

Of course producing mechanical defences is costly to the plant, because resources must be diverted away from producing leaves and the ones that are produced photosynthesize less efficiently. Some plants only produce protected leaves where they are required. In savannahs it is only the lower leaves of *Acacias* that are protected from browsing giraffes by thorns. The upper leaves, which are out of reach, are bare. Similar **heterophylly** may also be seen in holly bushes. *Ephedra* (Fig. 9.7), a member of the gnetales, does not even have leaves, but only particularly tough photosynthetic twigs.

9.7 ECOLOGY OF HIGHLY SEASONAL AREAS

9.7.1 Plant distribution

Despite their similarities, there are marked differences between the vegetation of savannahs, prairies, maquis and heaths. At least to some extent these differences are caused by the contrasting climates and soil conditions of the different regions. The savannahs and prairies, with their relatively productive growing seasons, are dominated by grasses and other perennial plants. In contrast, the drier Mediterranean maquis and some tundras are dominated by shrubs for the same reason that wetter Mediterranean re-gions are dominated by evergreen trees; the growing season is too short to allow any perennials except the bulbed monocots to develop new shoots that can reproduce.

For similar reasons many heaths and upland moors are also dominated by the slow-growing shrubs which are the only plants that can survive on their extremely nutrient-poor and acidic soils. Members of the heather family, Ericaceae, are particularly suc-

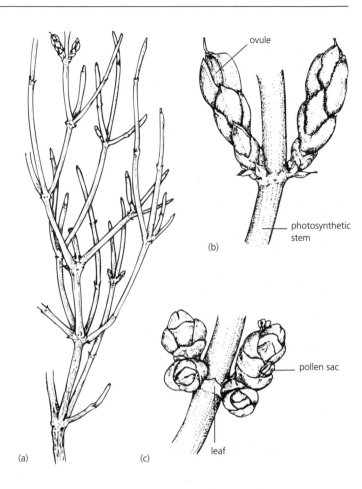

Fig. 9.7 *Ephedra antiryphilitica*, a member of a group of the gnetales common in dry Mediterranean areas. (a) Plan view (×0.8) to show how it has reduced its leaves to tiny scales. (b, c) close-up (×3.0) of petal-less female (b) and male (c) reproductive structures.

cessful in these regions, possibly because they have a mutualistic mycorrhizal association. The hyphae of these fungi ramify far into the soil and, in return for carbohydrates, greatly increase the uptake of nitrogen to the roots of the plant.

9.7.2 Succession

Unlike forests, many grasslands and heaths do not have stable vegetation; because they are dominated by plants of early succession, they tend to be open to invasion by trees. However, this is prevented by fires which destroy the invading trees. Herbivory is also an important factor, a good example being the effect of elephants on the tropical savannahs. In the course of their foraging, elephants often uproot and destroy invading trees, so maintaining their preferred grassland habitat.

9.8 THE EFFECT OF HUMANS

Humans have had, and are having, a much more dramatic effect on the vegetation of seasonal regions than have elephants! The major cause has been the clearing of forests for agriculture. As a result there has been a massive increase in the area of land which is artificially kept in an early stage of succession. In many cases this effect can be benign. Many cool temperate regions which would naturally be forested have been kept clear for thousands of years by simple methods such as grazing and burning. The species-rich **meadows** of lowland Europe and the Alps are such an artificial landscape. So too are the **moorlands** of upland Britain which have been kept free of trees by combined use of sheep grazing and periodic burning. Burning was also the tool used by native

North American Indians to maintain some of the wetter prairies in an open state which allowed them to hunt for bison. Similar methods are still used by nomadic hunters in the African savannah and Asian steppes.

Damage to the soil usually only occurs when the system is overexploited by excessive grazing or by agronomy. The soils of moorland Britain were degraded because of the agricultural activities of neolithic wheat growers. In such a condition they are now vulnerable to overgrazing by sheep or physical disturbance by walkers. Much more serious damage can happen to soils in hotter drier areas. In central Spain and in the prairies of North America, agriculture has led to a massive loss of topsoil and consequent production of 'dustbowl' conditions. As we shall see in the next chapter, deserts are also on the advance in many tropical and subtropical areas.

9.9 POINTS FOR DISCUSSION

1 Why do you think no plant has a poison which is potent enough to kill all animals, or has mechanical defences which foil all predators?
2 What do you think would happen to the vegetation of temperate regions if humans were killed off by a nuclear war?
3 What effect do you think global warming would have on the distribution of forests, grassland and tundra?
4 Why are there so few woodland grasses?

PRACTICAL INVESTIGATION

Collect small pieces of the foliage from any conifers cultivated or growing in your area. Look carefully at them, feel them and finally cut thin sections across the leaves and see if you can guess from the characteristics you can observe whether they might be deciduous, like the larch, or evergreen, like the pines. If they are cultivated, see if you can predict where they might grow naturally, then look them up in a reference book and see if you are right.

FURTHER READING

Archibold, O.W. (1995) *Ecology of World Vegetation*. Kluwer Academic Publishers, Amsterdam.
Crawford, R.M.M. (1989) *Studies in Plant Survival*. Blackwell Scientific Publications, Oxford.
Dallman (1998). *Plant Life in the World's Mediterranean Climates: California, Chile, South Africa, Australia and the Mediterranean Basin*. Oxford University Press, Oxford.
Harborne, J.B. (1988) *Introduction to Ecological Biochemistry*. Academic Press, London.
Horn, H.S. (1974) *The Adaptive Geometry of Trees*. Princeton University Press, Princeton.
Thomas, P. (2000) *Trees: their Natural History*. Cambridge University Press, Cambridge.

CHAPTER 10

Life on the edge

10.1 INTRODUCTION

10.1.1 Extreme climates

Two very different parts of the world, **deserts** and **polar** regions, have climates that are extremely inimical to plant life not only seasonally but virtually all year round. Deserts are found at latitudes of 15–40 degrees where air that has been raised by convection at the equator, cools and falls to earth. As it does so, clouds disappear, so unlike the equatorial region, deserts have permanently clear skies. This affects their climate in two ways. First, it means that they have very low rainfall and any large storms occur only very rarely. Second, it increases the daily temperature variation; in the clear conditions the temperature can rise to over 40°C during the day but the lack of insulating clouds means it can fall to below freezing at night.

In some respects the polar regions are very different. The sun gives little warmth and does not even rise above the horizon for several months during the winter. As a result, temperatures are much lower, and can fall to below −60°C! On the other hand, polar regions are dominated, like deserts, by high pressure, have clear cloudless skies and are swept by dry winds. Therefore both habitats have wide temperature fluctuations and impose severe desiccation on plants.

10.1.2 Plant survival in extreme climates

Despite the difficulties, plants do live even in these most taxing environments. This chapter will examine the adaptations plants have evolved that enable them to survive and reproduce in these two contrasting,

but in some ways similar, habitats, and will discuss the reasons for the success of particular plant groups. It will become clear that in the most extreme conditions, angiosperms and the other groups of vascular plants are not necessarily very successful. The adaptations of other groups which might be considered more 'primitive' have allowed them to colonize and survive in inhospitable areas even more successfully.

10.2 LIFE IN THE DESERT

10.2.1 Survival strategies

The potential for plant growth in deserts is reduced by the need to withstand extreme lack of water, and, to a lesser extent, to cope with the wide temperature fluctuations. Plants have evolved three main strategies to overcome the scarcity of water. Drought-avoiding plants shut their metabolism down when water is scarce and survive in a dormant state. In contrast drought-enduring plants survive by improving their ability to take up and conserve what little water there is, while drought-resisting plants store water when it is plentiful for subsequent use during drier times.

10.2.2 Drought avoiders

Although their mean rainfall is very low, most deserts are subjected at long intervals to rainy periods. In extreme cases heavy storms occur during which many centimetres of rain can fall in a matter of a few hours. Drought-avoiding plants have adapted to grow and reproduce in the short periods of time following such rainfall, during which water is still plentiful. When the water runs out they simply shut down their activity until the next rain event.

Survival as spores or seeds

The easiest way for a plant to withstand drought is as a dormant spore or seed, covered by a protective coat. Indeed, spores and seeds can lie dormant in the sand for several years before rainwater triggers their germination. For this reason the majority of desert plants are extremely short-lived, and can germinate and reproduce extremely rapidly. Many desert angiosperms can flower and set seed only 6–8 weeks after germinating, although the record appears to be set by *Boerhaavia repens* of the Sahara desert which takes as little as 8 days! Hardy cyanobacteria also emerge from their thick-walled resting stages and 'bloom' on the sand.

For a short period after a rain event therefore the desert is a riot of colour as flowers compete for the pollinating attentions of insects, and the cyanobacteria turns the sand from yellow to blue–green.

Survival underground

Vascular plants can also survive periods of drought if they are herbaceous perennials which store energy in their underground storage organs and protect their meristems within perennating buds. Beneath the sand the storage organ and buds will be protected both from the desiccating effects of the sun and from herbivores. Some desert species store their energy in bulbs, like the crocuses of the Mediterranean, but the dominant perennials in deserts are grasses and sedges not unlike those which grow in seaside sand dunes. Rainstorms stimulate the underground organ to produce large numbers of shallow rootlets which ramify widely through the surface of the soil and take up as much water as possible. The plant also produces leaves which then supply energy for the plant not only to flower and set seed, but also to replenish the underground storage organ. Only when the soil dries out do the active parts wither and the plant become dormant again.

Because survival is so much easier underground some plants never emerge at all. 'Cryptoendolithic'

lichens, cyanobacteria and algae are able to live out their lives entirely within rocks. These tiny organisms occupy the tiny interstices between the particles of porous rocks, often several millimetres beneath the surface, where sufficient light nevertheless penetrates to fulfil their meagre needs. Occasional episodes of rain or moisture that allow them to grow out onto the rock surface and reproduce reveal them to be little different from their relatives that grow in wetter places—they just avoid drought by hiding from it.

Survival above ground

Despite the added difficulties, some plants *do* manage to survive above ground during prolonged drought. As we saw in Chapter 5, many lichens can close down their metabolism and survive complete dehydration. Some species can therefore survive on rock surfaces by not being 'alive' most of the time! These drought avoiders only metabolize perceptibly when moisture arrives, but in doing so they pave the way for other organisms. The acids they produce break down the rock, and their thalli hold moisture and, when dead, break down and provide the first nutrients other than the minerals in the rock. Over hundreds or even thousands of years they can create a substrate fit for more demanding organisms such as mosses and flowering plants.

Most of the vascular plants that are able to cash in on this only slightly less hostile desert scene lose their leaves in times of drought, so there are a large number of deciduous desert shrubs such as sagebrushes. However, two plants from very different taxonomic groups show similar extraordinary adaptations of their foliage. The 'resurrection fern' *Pleopteris polypodioides* is a powerful example of how difficult it is to generalize about the characteristics of particular groups of plants. As we saw in Chapter 6, the ferns require water in order to reproduce, and you could be forgiven for thinking that this would limit them to damp environments. The resurrection fern, however, survives quite happily in the blistering heat of the prairies of the United States, partly because its leaves can endure the loss of up to 76% of their moisture; most other plants die after losing less than 12%. On a hot dry day the plants look dead and contorted, like the ferns you can see on walls (see Plate 1c, facing

p. 110) but within hours of the next rain shower the leaves expand into healthy vibrant green upright fronds (see Plate 1d, facing p. 110), helped by water-absorbing trichomes on the underside of its fronds. The 'resurrection plant' *Myrothamnus flabellifolia*, an angiosperm from south-west Africa does much the same thing; its leaves are capable of surviving even after becoming so dry that they can be rubbed into fine dust between finger and thumb. However, recovery is rather slower; they take over 12 days to rehydrate and resume photosynthesis when they are rewetted.

10.2.3 Drought endurers

A second set of plants, the drought endurers, have a unique set of adaptations which allow them to maintain photosynthesis and grow even during the long periods of drought between rainstorms. They have adaptations that both reduce water loss and increase the uptake of the water that is present, albeit scarce, even during droughts.

Reducing water loss

The vascular plants have **xerophytic** leaves which possess adaptations that reduce water loss; they are small, with thick cuticle and sunken stomata and some are able to curl up in extreme drought. Many drought endurers have vertical leaves to minimize sun exposure and so reduce water loss while the leaves of others track the sun, keeping their leaves parallel to it and so reduce heat load.

Water uptake from deep in the soil

Although most drought-enduring plants use a similar set of techniques to conserve moisture, they use several different techniques to obtain it. Some plants, the so-called **phreatophytes** exploit the water resources which, even in deserts, are available at the water table, deep below the soil surface. They develop roots that can penetrate to depths of up to 50 m. These deep 'tap' roots are energetically expensive to produce but once they have reached this permanent water source they can supply even large plants with adequate water. Many phreatophytes,

such as the acacias and tamarisks of Africa and *Prosopis* of central America, therefore grow into trees. The main difficulty such plants have is in establishing themselves in the first place, because the roots have to penetrate long distances through bone-dry soils. This restricts phreatophytes to areas, such as dried-up river beds, where run-off of rain waters periodically wets the soil for relatively long periods.

Water uptake from shallow soil

Another group of plants obtains moisture from the relatively frequent light falls of rain to which some deserts are subject and the heavy falls of dew which occur towards dawn. Since this moisture does not penetrate far into the soil, it is best taken up by extensive but shallow root systems. The roots of the creosote plant for instance may extend for up 50 m from its trunk but seldom grow deeper than 20 cm. The lower the rainfall the larger the root system that is needed to survive, and therefore creosote bushes become progressively spread out as conditions become drier.

Water uptake above-ground

A final set of plants do not use roots to obtain their water at all, but trap water above ground. The coastal deserts of Chile and Peru are subjected to sea fogs which provide a much more plentiful water supply than the infrequent rains. This supply is used by plants as large as acacia trees; fog droplets condense on their relatively unmodified leaves and drip to the floor where they are absorbed by their roots. Desert lichens of the genus *Rocella* use a similar mechanism. They look, and are constructed, just like lichens of wetter places, but manage to tolerate extreme desiccation during the day. As the early morning fog rolls in they uncurl, just like the resurrection fern, cram in what photosynthesis they can before the heat of the day, then curl up and tough it out until the next morning.

Further inland the vegetation is dominated by plants that are even better adapted to absorb moisture. Amongst these is the bromeliad *Tillandsia paleacea*, a close relative of the epiphytic bromeliads we examined in Chapter 8. This plant does not have

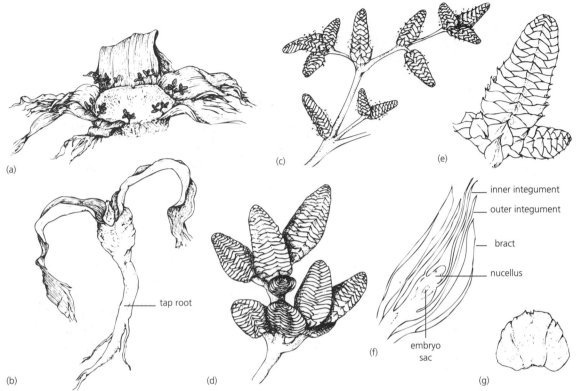

Fig. 10.1 The extraordinary desert gnetale *Welwitschia mirabilis.* (a, b) Views of an old and young plant, respectively (×0.2), showing the strap-like leaves which trap water from the fog, and the tap root which stores it. (c, d) Portions of the petal-less male (c) and female (d) reproductive branches (×1.0). The male flowers (e) have emerging anther-like structures, while in the female flowers (f) the ovule (seen here in longitudinal section) is protected by two integuments and a bract. The flower eventually produces a ripe winged seed (g).

roots, but survives in even the driest places entirely on the moisture it absorbs through the plate-like trichomes of its leaf surfaces (see Fig. 8.5, p. 141).

But the most extraordinary of all these plants is *Welwitschia mirabilis* (Fig. 10.1), a member of the gnetales, a small group of plants closely related to the angiosperms (see Chapter 7) and with unusual cone-like reproductive structures (Fig. 10.1, c–g). *Welwitschia* is endemic to the Namib Desert of Southern Africa, which, like Chile and Peru, is subject to coastal fogs. It produces just two strap-like leaves which emerge from a short trunk. These leaves collect droplets of dew which are channelled down grooves towards the centre of the plant. The water is eventually taken up and stored in a large conical tap

root. *Welwitschia* is an extremely slow-growing plant but since it never replaces its leaves, they may eventually grow to a length of 20 m and to an age of over 1000 years.

10.2.4 Drought resisters

The best known of all the desert plants are the **succulents**, such as the cacti and agaves. Their basic survival strategy is to take up more water than they need during wet periods and use their water store to maintain photosynthesis during periods of drought. They have a range of adaptations which speed up water uptake, increase the size of this water store, and prevent too much water being lost in transpiration.

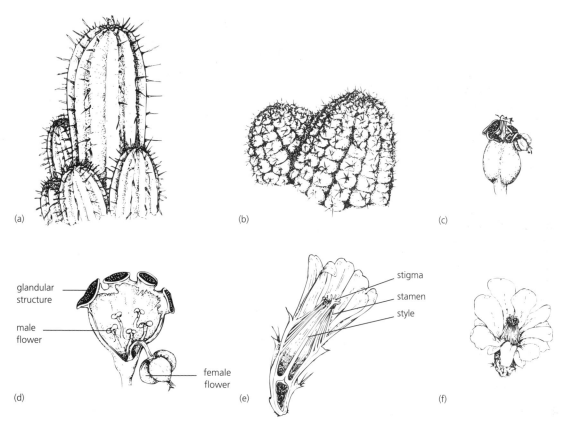

(a) (b) (c)

glandular
structure

male
flower

female
flower

(d)

stigma
stamen
style

(e) (f)

Fig. 10.2 Convergence in two desert stem succulents. The *Euphorbia* (a) from Africa and *Echinocereus* (b) a cactus from South America (both ×0.3) have similar overall form, having barrel-shaped stems, with spines at the apex of the longitudinal pleats. The differences between these members of different families is more apparent in their reproductive structures. *Euphorbia* (c) (×1.0) has an unusual inflorescence (cyathium) consisting of a cup made from four bracts each with its own glandular structure to attract flies. When cut open (d) this reveals a stalked female flower and several male flowers consisting of single anthers. In contrast *Eucactus* (e, f) (both ×1.0) has a single tubular flower with many stamens and a single style.

Root systems

Contrary to most people's preconceptions, most succulents do not have deep root systems but possess only a few shallow permanent roots. These resemble the roots of epiphytic orchids (Chapter 8) in possessing a **velamen** to reduce water loss. When it rains, though, a fine network of temporary roots is produced which absorbs water rapidly before the soil dries out again, after which they wither and die.

Water stores

The water that the roots absorb may be stored in a variety of structures. Leaf succulents such as agaves store water in a rosette of thickened leaves. This gives the leaves a large volume without increasing the surface area from which water can evaporate. Other species, such as the pachypodiums of South Africa and Madagascar, possess only a few, small leaves which can be lost in extreme drought. These plants store water instead in swollen fleshy stems.

The trend of leaf reduction and stem modification is taken to its ultimate conclusion in the stem succulents (Fig. 10.2), whose leaves are reduced to mere vestiges, while photosynthesis is carried out by the green swollen stem. In many plants the stem is star shaped in cross-section and is able to swell up like bellows during rain, storing large amounts of additional water. During the dry season the stem can then gradually shrink again as water is lost. The spines and hairs, which protect the stem from herbivores and animals out for a drink, also help to reduce water loss by shielding the stem from direct sunlight and trapping moist air. Water loss is further reduced by the massive thermal inertia of the swollen stem. The temperature is kept more constant, which has the further advantage of preventing chilling damage during the cool desert nights.

Convergence in stem succulents

Stem succulents have evolved independently several times in deserts in different parts of the world. Members of three different families—the cacti (Cactaceae), from America; and the euphorbias and milkweeds (Euphorbiaceae and Asclepediaceae) from Africa—have all developed the same body plan (Fig. 10.2a, b). However, their very different flowers (Fig. 10.2c–f) bear witness to their quite separate ancestry, which shows that these similarities are a case of **convergent evolution** (see Chapter 1). Their bodies are formed in quite different ways. The spines of cacti are modified leaves whereas those of euphorbias are made up of the tips of branches.

Each family of succulents has also independently evolved crassulacean acid metabolism (CAM) which further reduces water loss. CAM plants take up carbon dioxide during the night and convert it into stores of malic acid in their fleshy tissue. During the day the malic acid is then converted back to carbon dioxide which is used for conventional photosynthesis. The advantage of this arrangement is that the stomata of CAM plants need only open at night and can be shut during the day, greatly reducing water loss. CAM photosynthesis is not only found in desert angiosperms; there are also CAM ferns and even quillworts. The lesson seems to be that advantageous adaptations like CAM photosynthesis and water-absorbing trichomes crossphylum barriers.

Together the water-conserving adaptations of desert succulents can reduce water losses to as little as 25 g per gram of carbon dioxide fixed. In contrast a typical temperate herb would lose around 500 g. Further improvements in water efficiency can only be achieved by further drastic adaptations of body form. Perhaps the most bizarre inhabitants of the desert are the pebble plants of the family Mesembranthemoideae (Fig. 10.3). The entire plant is buried beneath the soil surface, leaving only the expanded tips of the leaves exposed. These are covered in dark spots, which act as windows to funnel light down to the photosynthetic tissue below. These strange plants are thereby adapted to live in dry stone beds, the more so because the spots also help make the plants look more like pebbles, and so help camouflage them from herbivores.

10.3 LIFE IN THE SNOW AND ICE

10.3.1 Survival strategies

There are two main problems facing plants in polar regions: plants must survive the long periods without sun; and cope with low temperatures, often amid ice and snow. The latter means that even when the sun does rise above the horizon, there may be such a deep layer of snow covering the plants that very little light penetrates until it has melted. This means that the times when temperatures and light levels permit a positive carbon balance decrease progressively towards the poles. Plants have as wide a range of strategies for dealing with these problems as we have seen for survival in desert conditions, but again, the real experts are found amongst the non-flowering plants.

10.3.2 Life in subpolar regions

In sub-Arctic **tundra** regions there is enough light for plant growth but they have two major problems: extreme exposure and cold above ground; and cold beneath the surface, where the ground may remain permanently frozen for all but the top thin layer. Roots simply cannot penetrate this **permafrost**, so cannot provide anchorage or adequate water and minerals for tall flowering plants. Indeed, because the thin topsoil expands and contracts on the frequent occasions on which it freezes and thaws, there is a danger that roots will be broken or levered out of the ground. Many Arctic plants have roots that are

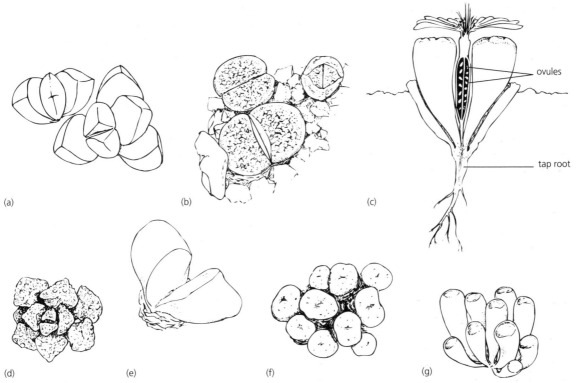

Fig. 10.3 Various 'living stones' (all ×1.0). (a) *Lapidaria*. (b) *Lithops*. (c) Vertical section through *Lithops* with a flower emerging from the central groove. Note that the ovules are protected within the centre of the plant. (d) *Titanopsis*. (e) *Argyroderma*. (f) *Conophytum*. (g) *Fenestraria* showing clearly the windows at the top.

curled around like helical springs which prevent such **frost heave**; such roots can easily stretch and contract with the surrounding soil.

Consequently, although many vascular plants do survive, tundra plants tend to be weird and wonderful versions of familiar temperate genera. Willows, probably most familiar as the tall, elegant riverside trees of temperate regions, are represented by dwarfed forms that eke out an existence by keeping their topmost parts buried by the snow in winter. Here willows grow less than 30 cm high, while birches may have just a single leaf.

We saw in the last chapter that the life-history strategy of plants depends on the conditions. This is also true in polar regions, although amongst the dwarf plants there is a range of strategies for coping with the harsh photosynthetic conditions. Members of the genus *Vaccinium* from the Ericaceae, for example, can be found as both deciduous and ever-

green forms side by side in sub-Arctic tundra. The deciduous strategy means that the plants can cut any winter losses through evaporation or damage sustained, but the cost is that of having to replace all the photosynthetic surfaces at the start of the new growing season. The evergreen strategy means that the leaves are all ready to go as soon as the snow melts; the cost is that of needing to protect and maintain photosynthetic structures that are acting only as wasteful respiratory centres during the winter. Paradoxically, snow cover acts to keep everything underneath it relatively warm, because it reduces heat losses through convection and infra-red radiation.

One of the best foliage designs for Arctic survival is that of the needle-like foliage of many conifers, such as the prostrate juniper bushes. The thick cuticles and sunken stomata of needles (see Fig. 7.6, p. 117), like those of the leaves of desert plants, are ideal for reducing water loss in the profoundly dehydrating winds of the sub-Arctic. Survival relies not just on structure, however, but also physiology. A process of 'hardening' accompanies the onset of winter, which makes the plants better able to tolerate the extreme conditions. As temperatures decline, the foliage becomes increasingly tolerant of freezing, and many conifers are able not only to tolerate temperatures well below freezing, but even to photosynthesize at these temperatures.

10.3.3 Life in the high polar regions

Guerrilla lycopods

Travelling towards the poles, where soils are thin, low in minerals and frozen for most of the year, angiosperms and conifers gradually give way to less demanding plant forms. In Chapter 6 we introduced the lycopod guerrillas: plants that can scramble across rocks and barren areas and only put down roots where tiny patches of soil are available. In the tundra another attribute of these highly successful plants also becomes important—that of highly efficient mineral recycling. Once taken up by the plant, minerals are held onto with incredible efficiency. Roots live far longer than those of flowering plants, up to 12 years or more, and remain physiologically active for almost half that time. Shoots are similarly tough, and increments of growth can often be traced back to intact stems over 20 years old. When these tissues finally senesce, the minerals, especially phosphorus, which is very limited in Arctic environments, are not lost into the environment but mobilized and recycled up to the new growing parts. This is one of the characteristics that allows lycopods to form vast mats, dominating the flora, in places where the low temperature limits microbial degradation of plant litter.

Mosses

No matter how efficient they are, the lycopods are limited to places where they can put down roots and take up the few minerals they need at least at some point in the year. No vascular plants can survive when the substrate remains permanently frozen. As we saw in Chapter 5, however, many mosses have no need of soil because they can obtain minerals or water from precipitation and can photosynthesize when the ground beneath them is frozen solid. They can therefore photosynthesize earlier in the summer season and outcompete vascular plants even where there is soil.

A good example of an Arctic moss is the feather moss, *Hylocomium*, which grows only at the tips of its branches, adding tiny increments of feathery growth to the modest, scrambling ground cover that it forms all around the northern hemisphere. Even on the bare rocks of many Arctic habitats, where no flowering plants can survive, this moss is still able to fuel the exceptionally brief growing season each year using rain alone.

Lichens

Lichens are even better suited to polar life than mosses. Most are not harmed by freezing, even to temperatures near to absolute zero, and many can photosynthesize and grow while still underneath the snow. Several species have been shown to take up moisture from snow, and have *optimal* light and temperature levels for photosynthesis that would kill the majority of flowering plants. Some lichens can photosynthesize at less than −24°C and at light levels less than 1% of full sunlight. It is unfortunate that brief periods of summer snow melt, higher temperature and insolation cause dehydration and photoinhibition in such species; the odd sunny day can wreak havoc with the knife-edge existence of these plants.

One way lichens cope with the problem of dehydration is to adopt the avoidance strategy we observed in some desert survivors. The real experts are again the **cryptoendolithic** plants. In the rock-strewn valleys of the Antarctic there is no visible life on the surface. This is not surprising because such places combine extreme cold and aridity; air temperatures range from a high of 0°C to −15°C in the summer, dropping to nearer −60°C in the winter and relative humidity from only 16–75%. Beneath the surface of some rocks, however, lurk a plethora of tiny photosynthetic organisms. Again, their survival is owed to both structural and physiological adaptations.

The community is dominated by lichens that

might form respectable thalli on the surface of rocks but become diffuse endolithic material within porous rocks. Fungal hyphae and clusters of algal cells grow between and around the crystals of the rock, so that the lichen becomes embedded within the rock matrix, protected by a hard surface crust. Less porous rocks are colonized by similarly simplified growth forms which delve into the microscopic cracks and fissures, becoming **chasmoendolithic**. These lichens are able to take in water solely from what tiny amounts of water vapour there is in the atmosphere; they have to, because the little snow that falls mostly sublimes without melting or is blown away by the wind.

Algae

The photosynthetic portions of lichens, the algal cells, are usually the *lowermost* layer of the structures. This is quite the reverse of what one might expect for associations requiring light in order to survive, and the opposite of most other lichens (see Fig. 5.13, p. 83). The uppermost layers are usually deeply pigmented, absorbing heat and so helping to ameliorate the low temperatures and also protecting the delicate photosynthetic cells from the high insolation of summer. Such protection from potentially damaging light levels is also thought to be the explanation for the deep red colour of the chlorophyte algae such as *Chlamydomonas nivalis*, which, in perhaps the ultimate achievement of polar plants, can live even on the surface of permanent snow.

10.4 POINTS FOR DISCUSSION

1 What are (a) the similarities and (b) the differences between desert and polar lichens?
2 Why are there no (a) short-lived annual plants or (b) water-storing plants in polar regions?
3 Why are there so few conifers in deserts?

PRACTICAL INVESTIGATIONS

1 Collect a few of the plants that are plentiful from very dry places in your neighbourhood. The tops of walls are a good place to look. See if you can tell how the plants are able to survive. For example, do the lichens look the same on the most exposed surfaces and in the shadier spots? See if you can find any on exposed dead branches of trees: are they the same colour on the tops as on the underneath of the branches? If you can keep returning to the same place, see what happens to the lichen if you remove the branch and reattach it to the tree upside down.

How do the mosses dry and rehydrate? Do they just close up their leaves or are they the real experts that curl their leaves around the stems and make a tight seal to prevent water loss? Peel off the leaves of the mosses and see where there are thick-walled empty cells, all ready to suck up moisture when it arrives. Are they on the periphery? The tips? The underside? How does their location relate to the way the moss behaves as it dehydrates?

Find out how the vascular plants lose and gain moisture. You should be able to see stomata and trichomes under a dissecting microscope, so look at the leaves and raise hypotheses to test. If you find a real drought endurer, such as rusty-backed fern (*Ceterach officinarum*), you may find stomata are restricted to the undersurface, and spot many elaborate trichomes on the underside too. You might predict that the fern would lose most of its moisture through the underside and regain water through its roots. You could test this by putting petroleum jelly on one, both, or neither surface of detached fully-hydrated leaves and weighing them. The jelly will stop evaporation. Put the leaves in a desiccator and weigh them every day until you can decide which surface loses most water. Then take dehydrated plants and put water on the leaves (top or bottom surface) and on the roots of different plants. Which hydrate fastest? Why might this fern curl up to present the **undersurface** of its leaves to the atmosphere rather than the topmost?

Many shops sell 'air-plants' (*Tillandsia* spp.)—but can they really 'absorb moisture simply from the air'? Remove leaves and suspend them in chambers with different levels of humidity (drops vs. large dishes of water in small sealed chambers) and see if their weight increases or decreases. If you put a drop of water onto the leaf does it roll off, soak straight in or get pulled along the leaf by the trichomes? See if you can see how the trichomes move as the leaf takes up water.

2 We have stated that the stems of barrel cacti swell up when supplied with water. Test how much the plants can swell up by taking measurements of the stem before and after several days of heavy watering, weighing before and after. What is the ultimate limit to the stem diameter? Cut open the cacti and look inside. Where is the water held? Some cacti can grow over 10 m tall. Where do they get the strength to stand up when they are drought-stressed?

FURTHER READING

Archibold, O.W. (1995) *Ecology of World Vegetation*. Kluwer Academic Publishers, Amsterdam.

Crawford, R.M.M. (1989) *Studies in Plant Survival*. Blackwell Scientific Publications, Oxford.

Fogg (1999) *The Biology of Polar Habitats*. Oxford University Press, Oxford.

Nobel, P.S. (1988) *Environmental Biology of Agaves and Cacti*. Cambridge University Press, Cambridge

CHAPTER 11

Life back in the water

11.1 INTRODUCTION

11.1.1 Problems of life in mud

Despite their adaptations to life in water, the algae we examined in Chapter 4 are unable to colonize the muddy bottoms of rivers, lakes and estuaries. Simple encrusting algae would swiftly be covered by mud, while the more complex filamentous and parenchymatous algae have no sufficiently solid substrate to which they can glue their holdfasts. Paradoxically this niche is instead dominated by those very plants whose adaptations to dry conditions we have examined over the last four chapters: the vascular plants and in particular the angiosperms! Indeed, as we shall see, it is one of the very adaptations which so suited vascular plants for life on dry land—the possession of a root system—which has proved to be the most important preadaptation to life back in the water.

Undoubtedly the route for plants readapting to life in the water was via the waterlogged soils of marshes and around the banks of water courses. In some ways waterlogged soils are ideal habitats for the growth of vascular plants because they provide them with a constant supply of water. However, the water also has a negative effect because it displaces air from the spaces between the soil particles and so stops oxygen diffusing to the roots. Any oxygen that does diffuse down through the water is quickly used up by bacteria and other soil microorganisms and never reaches the roots. In these anaerobic conditions the root systems of ordinary land plants are unable to respire and quickly die (a good reason not to over-water your house-plants!). Some roots can survive for a time by using the much less efficient anaerobic respiration, but even they will eventually die. Ethanol and other metabolites will build up and poison the root tissue.

11.1.2 Adaptations for waterlogged conditions

Pneumatophores

The key to surviving in waterlogged conditions is to pipe oxygen down to the roots, and waste gases away from them. One method, which has been independently developed by several groups of trees, is for the roots to grow special breathing organs called 'knee roots' or **pneumatophores** (Fig. 11.1). These are hollow upward extensions of the roots which are linked to the outside air by lenticels. These extensions are used by the swamp cyprus *Taxodium distichum*, which forms extensive forests in the flooded areas of Florida and Georgia, such as the famous Everglades.

Root aerenchyma

However, the most common method used is one that requires only a very small change in morphology. Vascular plants need simply extend the air spaces which they all contain within their leaves and stems down into their roots. This **aerenchyma** acts essentially like a supply pipe, linking roots via the stomata to the outside air. Many flood-tolerant trees, such as the lodgepole pine *Pinus contorta*, produce aerenchyma by enlarging the intercellular spaces within the stele of the roots. Most flood-tolerant herbs, in contrast, such as the horsetails, sedges, reeds and rushes which dominate the shores of most rivers and lakes, achieve the same effect by enlarging the intercellular spaces within the root cortex (Fig. 11.1a, b).

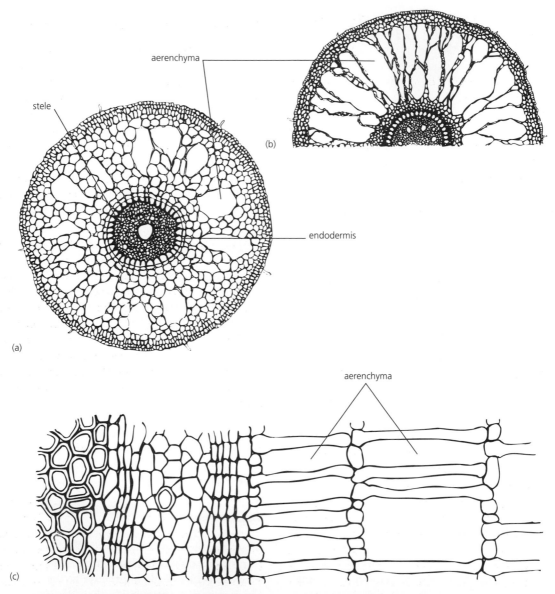

Fig. 11.1 Transverse sections through the roots of semi-submerged vegetation showing the **aerenchyma** which pipes air to the respiring structures. (a) The reed *Phragmites communis* and (b) the rush *Juncus effusus* (both ×10). (c) Portion of a transverse section of a **pneumatophore** of the tropical mangrove *Jussaea peruviana* (×200).

Stem adaptations

Extending air spaces down from the shoots to the roots also allows plants to live in standing water. The lower region of the stem is supplied with air just like the roots and can survive submersion equally well.

Many of the monocots, such as rice, which survive flooding so well have hollow stems down which gases can diffuse unimpeded to the roots. The leaves of rushes are even better adapted for air movement; they not only have air spaces within them but also curl around to form a much wider air-filled pipe.

They therefore resemble the stems of the horsetails we examined in Chapter 6.

But there is another advantage of extending air spaces within the entire plant. It links the parts of the plant that produce oxygen (the leaves) with the plants that use it (the roots) and so allows plants to pipe the oxygen they themselves produce to where it is most needed. In certain unusual habitats there may be a further advantage to having aerenchyma. The quillwort *Isoetes histrix*, for example, survives well in bogs partly because its air spaces transport the carbon dioxide produced by decomposition in the soil up through the roots to the leaves. This unique supply system provides the plant with over half the carbon dioxide it needs for photosynthesis.

11.2 LIFE IN FRESH WATER

11.2.1 Strategies for survival

Aerenchyma is the key to plant survival in both waterlogged soil and fresh water. However, the morphology of water plants varies greatly depending both on the depth of water they live in and the speed at which it is flowing. There are essentially four strategies water plants can use for survival. In shallow and slow-moving water, plants have emergent shoot systems similar to those of land plants.

In deeper and faster-moving water the stems of such emergent plants would be unable to grow to the surface. Indeed, even if they did so they would break under the weight of the crown and the large drag forces to which they would be subjected. In these conditions plants use one of three alternative strategies. The first is to develop floating leaves which need no support from the stem. In still deeper and faster flowing water, however, even plants with floating leaves would be unable to survive. Their stems may not be able to reach the surface or would snap under the force of the water. In these areas they are replaced by free-floating plants or plants which have lost the connection with the surface and become submerged. As we shall see, plants which have chosen each of these strategies are subjected to quite different selection pressures and so differ greatly in form.

11.2.2 Plants with emergent shoot systems

Plants with most of their shoot systems above the water surface need very few adaptations to aquatic life apart from the need for aerenchyma. The aerial shoots can photosynthesize normally and reproduce just like those of land plants. They do show a few other differences, however. Most of them possess larger and less dissected leaves than their relatives on dry land (see Fig. 11.7f). Normally such large leaves would overheat in the sun, but because aquatic plants have a constant water supply they can keep cool by allowing rapid transpiration through their stomata.

Many aquatic plants also have seeds which are dispersed by water, the medium which is most likely to take them to suitable germination sites. These seeds are usually buoyant, having layers of air-filled tissue and a water-repellent coat. They are released during dry periods, when they can wash up and germinate on the shore in areas which are usually underwater. The seeds of the water violet *Hottonia palustris* actually germinate immediately they are shed and the seedling establishes itself as soon as it reaches dry land.

11.2.3 Plants with floating leaves

Basic leaf morphology

Water plants with floating leaves (Fig. 11.2) differ much more even from related land plants. To stay on the surface their leaves develop extensive aerenchyma to provide buoyancy. Their stomata are also restricted to the upper surface of the leaf rather than the lower one, and this surface is kept dry by a waxy water-repelling cuticle. The lower surface has no stomata but in some species it does have an absorption function. In the water lily *Nymphaea* it is provided with groups of **hydropoten cells** which take up ions from the water. The modified leaves are linked to the root system by elongated petioles. Like the leaves, these petioles also contain extensive aerenchyma which both provides buoyancy and aerates the root system. Apart from this, however, the petioles are not greatly modified from those of land plants; they are longer and more flexible

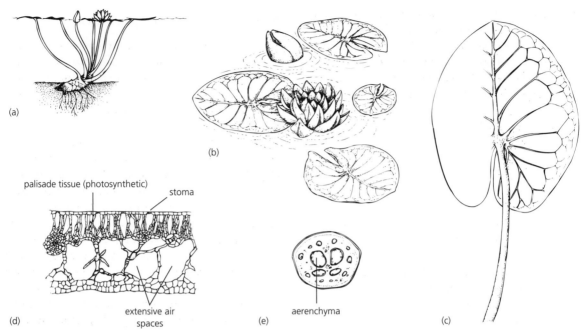

Fig. 11.2 Adaptations to life floating on still water in the water lily *Nymphaea*. (a) Growth form showing the long petioles radiating out from an underground rhizome. (b) The plant seen from above (×0.1). (c) The leaf, showing its almost circular shape. (d) Vertical section of the leaf (×10). Note the presence of stomata on the upper surface only and the extensive air cavities within the mesophyll which provide buoyancy. (e) Transverse section of the petiole (×2.0), showing the aerenchyma which pipes oxygen down from the surface to the roots below.

but they do contain well-developed xylem to transport water from the roots up to the transpiring leaves.

Adaptations to still water

Plants with floating leaves all tend to share the same basic design but plants which grow in different environments do show particular morphological adaptations. Still lakes and slow-flowing rivers are dominated by plants possessing circular leaves (Fig. 11.2c) which most efficiently compete for space. Typical examples are the well-known water lilies which grow in a rosette form (Fig. 11.2a, b). As new leaves are pushed up in the centre of the rosette, the petioles of the older leaves carry on elongating, allowing them to move outwards to give the youngsters room and enabling the plant to take over more of the water surface.

Of course, with so many plants competing for limited space the leaves of each plant are in danger of being overgrown by those of its neighbour and being shaded out. The huge South American water lily *Victoria amazonica* (see Plate 2c, facing p. 110) protects itself from this possibility by developing upturned leaf margins. The leaves of this amazing plant, which may reach 2 m in diameter, are structural masterpieces. They are supported by a series of air-filled veins which provide enough support and buoyancy for leaves to withstand the weight of a small child. Rainwater is drained away from the leaf surface through small holes. So impressive is the leaf structure that it is said it was used by the architect and gardener Joseph Paxton as the inspiration for his design for his huge 'Crystal Palace' that housed the 1851 exhibition in Hyde Park, London.

Adaptations to fast-flowing water

In faster-flowing rivers and streams circular leaves are at a disadvantage because the upstream lobe may be bent over by the pressure of water flow. Here, plants with circular leaves are replaced by species with long, thin ones which present much less resistance to the flow. This trend is shown clearly by the pondweeds of the genus *Potamogeton*. The leaves of these plants also possess two other adaptations to resist tearing: they have smooth outer margins, to prevent the outer edge from tearing; and are reinforced by longitudinal veins which deflect any tears which do form.

11.2.4 Free-floating plants

Basic morphology

Free-floating plants are found throughout the world and include representatives of two main groups of plants: the angiosperms and the ferns. The least derived of the free-floating plants, such as the water hyacinth *Eichornia crassipes*, have a rosette of buoyant leaves like those of water lilies, but, unlike them, it is attached directly to the roots.

Nutrient uptake

Water contains lower concentrations of nutrients than a typical soil, so the fibrous roots of *Eichornia* are covered in a profusion of long root hairs which greatly increase the area for absorption. A similar situation can be seen in the floating fern *Salvinia* (see Plate 2, facing p. 110). Even so, it is difficult for free-floating plants to obtain adequate nutrients, so they are restricted to water courses which are rich in dissolved salts. *Eichornia* reproduces largely asexually by producing new individuals on the ends of stolons, and these tangle up together, helping the plants carpet not only on lakes but also slow-moving rivers. It has recently been widely introduced throughout the tropics and has become a serious pest in Lake Victoria.

The problem of obtaining an adequate supply of nutrients is solved in a different way by the small floating dicot *Aldovandra vesiculosa* (Fig. 11.3a). Like its terrestrial relative the Venus fly trap *Dionea*, it traps insects within its jointed leaves (Fig. 11.3b), closing them rapidly when an insect touches sensitive hairs on its surface. The insect is then digested by secretions from glands on the surface of the leaf and its body fluids absorbed.

Morphological reduction in floating plants

Although *Eichornia* is extremely successful, there is no real need for floating plants to have a complex structure because they have little requirement for transport or structural strength. Floating plants therefore, tend to show reduction in both size and complexity.

Two genera of water ferns are particularly famous and show this trend clearly. One is *Salvinia molesta* (Fig. 11.4, Plate 2a, facing p. 110), whose common name, Kariba weed, reflects its ability to block waterways and dam constructions. This fern has reduced vegetative features which allow it to double its dry weight in $2\frac{1}{2}$ days. Its horizontal stems bear no roots, only highly modified leaves, the upper ones being simple in shape but being coated by elegant waxy hairs. Complex water-repellent hairs, the so-called **egg-beater trichomes** which cover the leaves (Fig. 11.4c, Plate 2b, facing p. 110), allow the plant to bob back up to the surface if swept under. The lowermost leaves replace the anchoring and absorptive function of roots, which indeed they much resemble.

A benign relative of *Salvinia* is *Azolla* (from the Vietnamese 'to dry, to die'), a floating fern every rice farmer loves. This fern is even more reduced, being composed of a simple shoot system which looks rather like a leafy liverwort. The tiny leaves are arranged in two ranks on either side of the horizontal stem. Each leaf has two lobes: the photosynthetic upper lobe is buoyed up by a covering of epidermal hairs; the colourless lower lobe is submerged and absorbs dissolved nutrients.

A cavity in the upper lobe of each leaf contains filaments of the cyanobacterium *Anabaena azollae* which fix atmospheric nitrogen. The cyanobacteria, as is often the case in mutualistic relationships, are highly modified by the association, and form few cells which are not heterocysts (see Chapter 3). The nitrogen that is fixed is welcomed by the fern (which contributes fixed carbon to the cyanobacterium) but also by the rice farmers. The ferns thrive in wet paddy fields, fixing nitrogen which is given up to the rice plants when the paddy fields dry out later in the season. Farmers have known this for centuries, and temples dedicated to *Azolla* serve to provide one

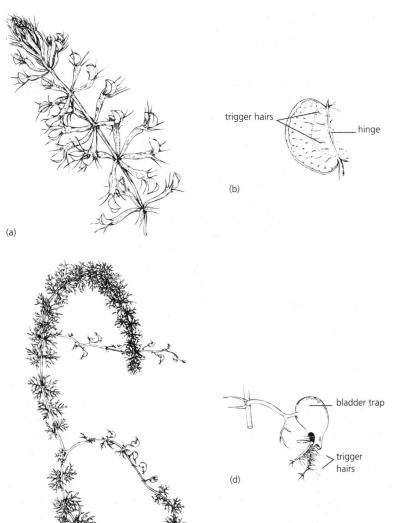

trigger hairs

hinge

(b)

(a)

bladder trap

trigger hairs

(d)

(c)

Fig. 11.3 Carnivorous aquatic plants. The floating dicot *Aldovandra vesiculosa* (a) (×2.0) traps insects within jointed leaves (b) (×6.0) just like its close relative the Venus fly trap. The bladderwort *Utricularia intermedia* (c) (×0.8) traps small aquatic crustaceans like *Daphnia* within its submerged traps (d) (×8.0). When the trigger hairs are touched the trap opens and the insect is sucked in.

cool damp environment where the fern can be kept alive until the next season.

However, the greatest reduction in size and complexity is seen in the monocotyledonous family Lemnaceae (Fig. 11.5). The common duckweed *Lemna minor* (Fig. 11.5b), for instance has a shoot system which is reduced to a single leaf-like thallus a few millimetres in diameter, and just a single root. Some members of the genus *Wolffia* (Fig. 11.5e, f) even lack roots or a vascular system and resemble floating buttons.

11.2.5 Problems for submerged plants

Plants whose leaves are totally submerged beneath the water have greater problems to overcome than those with aerial or floating leaves. First, they have no access to aerial carbon dioxide for photosynthesis and must obtain it from the water, through which diffusion is much slower. Second, without water evaporating from their leaves there is no power source to drive the transpiration stream and so draw up nutrients from the roots to the shoot. Finally, the

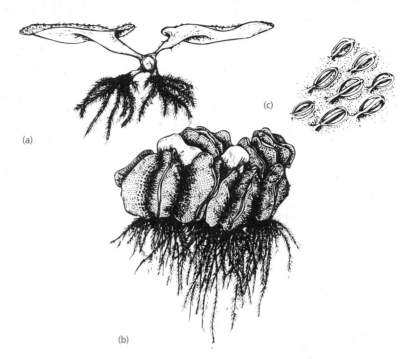

Fig. 11.4 The floating fern *Salvinia molesta*. (a, b) Side and general views (×1.0), showing the buoyant hairy upper leaves and the profusion of hairy root-like lower leaves. (c) Close up (×10) of the water repelling egg-beater trichomes on the upper surface of the leaf which helps maintain buoyancy and keeps the plant upright.

plant may be unable to raise even the flower shoot out of the water to allow wind or insect pollination to take place. All of these difficulties have driven some submerged plants to modify their morphology to such an unprecedented extent that they entirely lose their resemblance to their terrestrial relatives. However, still and fast-flowing waters exert quite different selection pressures on the plants which inhabit them and this has resulted in a marked divergence in their morphology.

11.2.6 Submerged plants of still water

Morphological adaptations of the shoot system

In deep still water, photosynthesis will be particularly seriously limited by the supply of carbon dioxide to the leaves because they must rely on its extremely slow diffusion through water. One way of reducing this problem, which is seen in the submerged water weed *Potamogeton pectinatus*, is to develop a rosette of narrow, strap-shaped leaves. Alternatively narrow leaves can be arranged in whorls around a central stem, as in the Canadian pondweed *Elodea canadiensis* (Fig. 11.6f).

An even more effective solution, seen in such plants as the hornwort *Ceratophyllum demersum* (Fig. 11.6a) and the bladderwort *Utricularia vulgaris* (see Fig. 11.3c) is to produce finely dissected leaves in which the webbing between the veins has been lost. By this apparent reversal of the original evolution of the megaphylls, the plants further increase the surface area of their leaves. This speeds up the supply of carbon dioxide and hence the rate of photosynthesis. To speed up photosynthesis yet further, the chloroplasts of most submerged plants are located in the epidermal tissue, where the supply of light and carbon dioxide are greatest. In cross-section the leaves therefore resemble at least at first glance the thalli of many brown seaweeds (Chapter 4). Some of the oxygen that is produced by photosynthesis is retained between the cells of the plant just as in other water plants, and is transported, via the aerenchyma, to the roots.

Morphological adaptations of the root system

The morphology of these submerged plants also differs in other ways from their emergent relatives. Because they have no transpiration current and

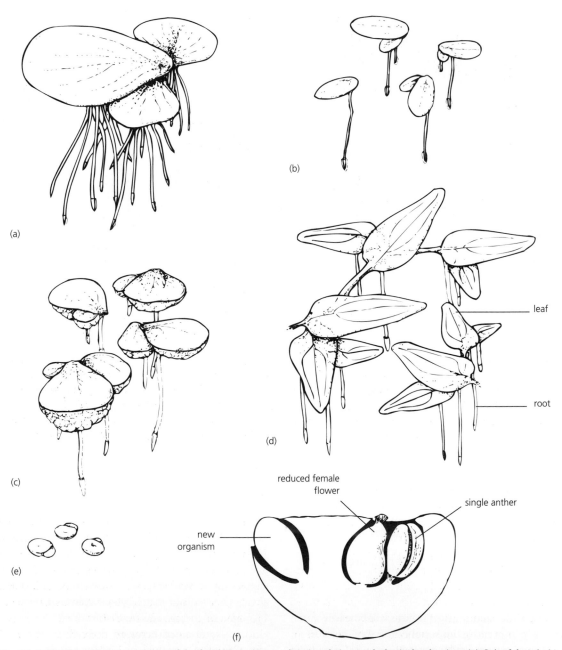

Fig. 11.5 Floating aquatic monocots of the family Lemnaceae, showing their morphological reduction. (a) *Spirodela polyrhiza* has leaves and several roots. In the genus *Lemna* the number of roots is reduced to one per leaf. (b) *L. minor*, (c) *L. gibba*, (d) *L. frisulea*. *Wolffia arrhiza* (e) has no roots at all (all ×6.0). A vertical section (f) showing the reduced flowers and the development of a new organism by asexual budding.

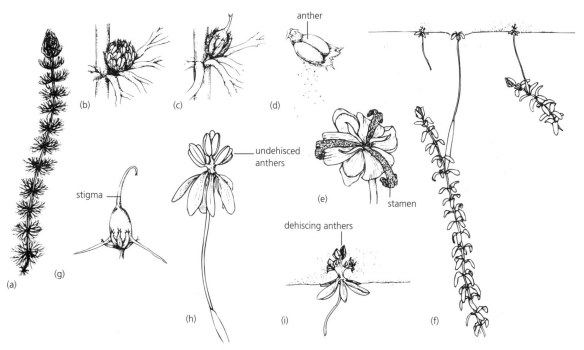

Fig. 11.6 Morphological and reproductive adaptations to life submerged in still water. *Ceratophyllum demersum* (a) (×0.3) has finely dissected leaves which improve carbon dioxide uptake. Pollination is underwater. The male flowers (b) are held above the female flowers (c) on the plant. Pollination is via the water when detached male flowers (d) release pollen and it falls onto the stamen of the female flower below. (e) A maturing fruit (all ×3.0). *Elodea canadiensis* (f) (×0.3) has less finely dissected leaves and pollination is along the water surface. Both the female (g) and male (h) flowers (both ×3.0) are held at the surface of the water by surface tension, attached or unattached. The male flower dehisces (i) to release hairy pollen grains which float along the surface to the female flower.

because the shoot system is supported by the buoyancy of the aerenchyma, there is no need for extensive xylem, and it is reduced. However, without a transpiration current it becomes hard to supply nutrients from the root system up to the shoot system. For this reason, the root systems of submerged plants have mainly an anchorage role and are reduced in size. Nutrients are instead increasingly taken up through the epidermis of the dissected leaves. Such is the reduction in importance of the root system that species such as *Ceratophyllum* and *Utricularia* dispense with roots altogether and float beneath the surface. Of course, as we have already seen, it is difficult for floating species to obtain adequate nutrients from water. *Utricularia*, like *Aldovandra*, overcomes this problem by resorting to carnivory, trapping unwary crustaceans in hollow bladder-like traps (see Fig. 11.3d).

Pollination

Despite having a submerged vegetative body, most species that live in still water do manage to raise the flower head out of the water and make use of insect or wind pollination. The flower stalks of several species of *Utricularia* and *Ranunculus*, for instance, are held aloft by whorls of modified leaves which are inflated to form floats. A few species, however, do use water as a vector to carry their pollen. In the Canadian pondweed *Elodea canadiensis* (Fig. 11.6f–i) the separate male and female flowers are each held at the surface by water-repellent hairs. The pollen grains, which are themselves coated with such hairs, are released onto the water surface. They then float to the stigma of the female flower. True water pollination, occurring below the surface, occurs in only a very few plants of fresh water of which

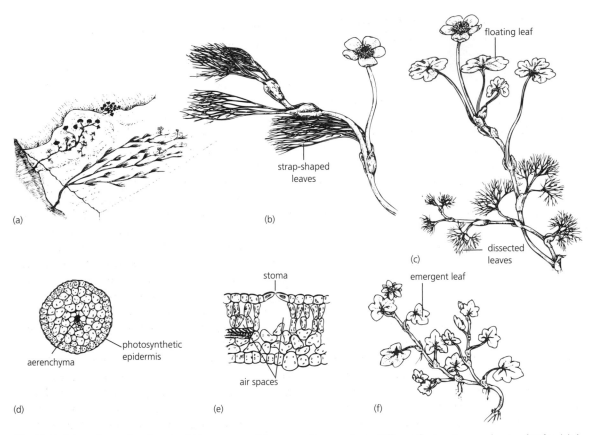

Fig. 11.7 Adaptation and acclimation in buttercups of the genus *Ranunculus* to different flow regimes and water depths. (a) A stylized view of a stream showing where different growth forms occur. *R. fluitans* (b) is adapted to life in fast-flowing areas, as its strap-shaped leaves flag downstream and are reinforced by central xylem strands. *R. peltatus* (c) is adapted to slow-moving water, and shows **heterophylly** (all plants ×1.0). Below the water it has finely dissected leaves which in section (d) (×15) show photosynthetic tissue only on the outer surface. However, it also has floating leaves which in vertical section (e) (×10) show the presence of stomata only on their upper surface and of air spaces which provide buoyancy. *R. hederaceus* (f) is a marginal species of shallow water which shows the characteristic broad leaves of emergent plants.

Ceratophyllum (Fig. 11.6d, e) is the best known. Like *Elodea*, *Ceratophyllum* has separate male and female flowers. Only the long-stalked male flowers reach the surface where they are held up by the buoyancy of their stamens. Once at the surface the pollen is released and rains downwards to the female flowers which are displayed below.

11.2.7 Submerged plants of fast-flowing water

Morphological adaptations

As we saw in Chapter 4, moving water supplies carbon dioxide and nutrients very rapidly to submerged plants. However, they must withstand large drag forces, and so mechanical considerations more strongly affect their morphology. Plants of fast-flowing streams such as *Ranunculus fluitans* (Fig. 11.7a, b) therefore possess long, strap-like leaves and rope-like stems, both of which are reinforced by strong, centrally placed strands of xylem. The stems flag out downstream with the flow and are anchored firmly to the bed of the stream by large numbers of strong adventitious roots. The great strength and flexibility of these plants allows them to survive even the spate flows which follow summer storms, and so high is the resistance to flow of large clumps of weed that they can cause the streams to flood.

Pollination

The main difficulty plants have in fast-flowing water is to carry out sexual reproduction. Water pollination is not possible because the flow will rapidly remove pollen grains downstream. In faster flow it is also harder to raise a flower stalk above the surface of the water. Many plants, like *Ranunculus*, make use of the buoyancy of their stems and use modified leaves as floats. However, *Myriophyllum* actually makes use of the flow. It raises its flower spike by so orienting its leaves like wings that they provide hydrodynamic lift!

The Podostemaceae

But perhaps the most extraordinary of all freshwater plants are the members of the tropical family Podostemaceae, most of which inhabit torrential tropical streams. These plants are anchored to bare rock by root systems which have been modified to form a creeping photosynthetic thallus. This is highly reminiscent in some cases of the bodies of liverworts and lichens, or in others of the holdfasts of marine algae; it creeps over the rocks and is cemented to them by hairs or exogenous projections called haptera. In the less specialized forms, shoot systems rather like those of other species of submerged plants are attached to the thallus. However, in many species the thallus is increasingly dominant and the shoot system much reduced. In *Dicraea stylosa* parts of the broad thallus float free in the stream, looking strikingly like that of the brown seaweed *Fucus*, while the shoots appear only at its tips and have no photosynthetic function. The similarity of these astonishing plants to other lower plants is more than skin deep. Without subterranean roots they have no need for an aeration system and they lack aerenchyma. Their xylem system is also reduced. The result is a solid body very like that of brown algae. As is so often the case in biology, similar selection pressures acting even on organisms which have very different ancestors has produced spectacular convergence in body form.

11.2.8 Environmental influences on the form of water plants

The description of the adaptations of water plants to different habitats may have given the impression that each species of water plant is capable of living in just a single habitat. But although species do show strong adaptations to particular conditions (see Fig. 11.7 for a description of adaptation in the genus *Ranunculus*), individual plants of a single species may be capable of acclimatizing to quite different habitats, and even a single plant may change its morphology over its development or as the conditions in which it lives alter. The water crowfoot, *Ranunculus peltatus* (Fig. 11.7c), is like *Salvinia* (see Fig. 11.4, Plate 4, p. 110) in showing **heterophylly,** producing feathery leaves underwater and relatively normal-shaped leaves which float on the water surface.

Similarly, the arrowhead *Sagittaria sagittifolia* is **polymorphic**, producing arrow-shaped leaves when grown in mud but only long, ribbon-shaped, submerged leaves when grown in very deep water. When grown in water of intermediate depth a single plant can also show heterophylly, producing three different forms of leaves: narrow, ribbon-shaped, submerged leaves; oval leaves floating on the surface; and arrow-shaped leaves above the water. The plant even shows a transition as it ages; fewer ribbon-shaped leaves and more oval and arrow-shaped ones are produced.

11.3 ADAPTATIONS TO LIFE IN SALT WATER

11.3.1 Strategies for survival

Although many vascular plants have readapted to life in rivers and lakes, far fewer have been able to colonize the mud which is deposited around the mouths of estuaries and gently sloping shores. This is largely because of the high salinity of sea water. Its high osmotic pressure tends to draw water out of land plants, which have adapted to the low concentrations of ions that are typically found in rainwater and in the soil. The removal of water would therefore cause them to wilt, while the high concentrations of sodium ions in sea water would tend to build up in the cells of the plants, and eventually poison them.

Quite drastic alterations in physiology and morphology are needed to stop these things happening. There are several ways of keeping down concentrations of sodium. Its entry can be blocked by carrying out ultrafiltration in the roots. Alternatively sodium can be actively excreted by special salt glands, just as in marine reptiles, or removed by concentrating it into tissues which are then jettisoned. Water loss may be reduced in two main ways: losses to the surrounding medium can be reduced by increasing the osmotic concentration of the cells; and transpirational water losses can be reduced by making the leaves smaller and thicker. As a result, despite being surrounded by water, many plants which have adapted to life in sea water look superficially very similar to the **xerophytic** and **succulent** plants of deserts.

The only group that has been able to make the necessary adaptations are the angiosperms, and there are many species of marine flowering plants. The strategies that these plants adopt depend on the temperature of the water in which they are growing as well as its depth. In the warm tropics the intertidal zone is dominated by **mangrove** trees which use ultrafiltration, while in temperate areas this zone is dominated by **salt-marsh** perennial herbs, which show greater morphological adaptations. Subtidal zones are dominated by the **seagrasses**, which resemble, at least superficially, submerged freshwater plants.

11.3.2 Mangroves

The shores of tropical estuaries are dominated by a number of species of evergreen trees which form **mangrove swamps**. Mangroves have evolved independently several times from quite unrelated ancestors and so include members of several different families of angiosperms. However, because they face similar selection pressures these plants show a wide range of convergent adaptations which has resulted in them looking remarkably similar.

Morphological adaptations

The leaves of mangroves are thick and leathery to reduce water loss. But the most spectacular adaptations are those which improve the anchorage and aeration of the root systems. Mangroves are anchored in the soft mud both by **prop roots** which emerge from their trunk and by **drop roots** which grow downwards from their lower branches. Both of these sorts of root contain extensive aerenchyma to draw oxygen downwards, and aeration is further improved by the knee-like **pneumatophores** (see Fig. 11.1c) which grow upwards from the lateral roots. Good root aeration is particularly important because oxygen is required to power the ultrafiltration process which excludes salt.

Reproduction and colonization

The flowers of mangroves, which are held well above the high tide mark show no particular adaptations to marine life, although because shorelines tend to produce sea breezes many mangroves are wind pollinated. However, many do have fruits specially adapted for colonization and water dispersal. The long thin fruits of *Kandelia*, for instance, are well adapted either to spear directly into the mud if dropped at low tide, or to float away if dropped at high tide. In this way mangroves rapidly colonize new areas of mud which then gradually builds up around the tangle of roots. Eventually the landward side of a mangrove swamp rises above the high tide mark and the mangroves are replaced by terrestrial species. Hence mangroves are a key part in the process of primary succession on tropical shores. The swamps also act as an excellent habitat for a wide range of fish and invertebrates. Unfortunately their richness is being exploited by humans. Many areas are being cut down to provide temporary areas for shrimp farming, and the destruction this causes is the biggest threat to these unique habitats.

11.3.3 Salt-marsh plants

Succulents

Mangroves seem to be unable to survive in places where the sea temperature falls to below about 20°C in winter, perhaps because they are then unable to filter out salt fast enough. They are replaced in

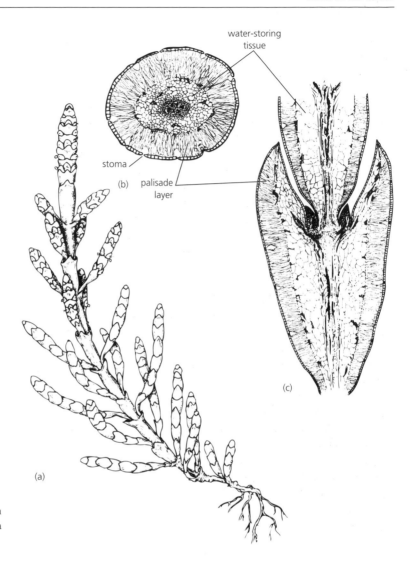

water-storing
tissue

stoma

(b)   palisade
layer

(c)

Fig. 11.8 The succulent salt-marsh herb *Salicornia fruticosa* (a) (×1.0). Transverse (b) and longitudinal (c) sections of the same plant showing the internal water-storing tissue beneath the palisade layer and the leaf reduction which produces a plant reminiscent of a spineless cactus.

(a)

temperate regions by a range of herbaceous and shrubby dicots and monocots which together form the vegetation of **salt marshes**. The dicots include annual **succulents** such as the saltworts *Salicornia* (Fig. 11.8) and *Sueda maritima* which colonize and help to stabilize lower areas of the marsh. This gradually raises its level and leads to their replacement by perennial species such as sea plantain *Plantago maritima* and shrubs such as the sea purslane *Halimione*.

Xerophytes

However, many salt marshes are dominated by perennial monocots, which have **xerophytic** rather than succulent leaves. Their resistance to salt is therefore conferred by their impermeability to salt uptake and to their ability to excrete salt through salt gland cells on the epidermis of their leaves. Lower areas of salt marshes around Northern Europe are colonized by the newly evolved grass *Spartina anglica* (see Chapter 1), which is replaced in upper areas by rushes such as *Juncus* and reeds such as *Phragmites*. Just as in fresh water, the hollow stems of these angiosperms are undoubtedly a preadaptation which allows them easily to develop aerenchyma to supply flooded regions with oxygen. They grow from

183

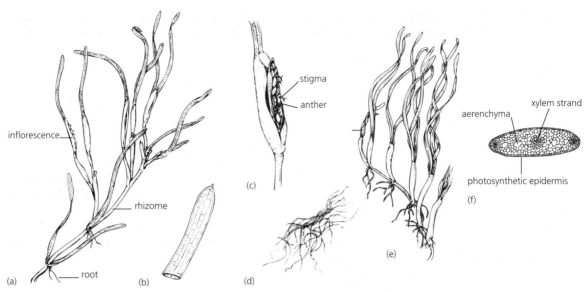

Fig. 11.9 Morphological and reproductive adaptations in seagrasses. *Zostera marina* (a) (×0.2) is typical in form, having ribbon-like leaves (b) emerging from a rhizome which is anchored by fleshy roots. The inflorescence (c) releases pollen grains which are joined together in gelatinous strands (d). *Halodule wrightii* (e) (×0.2) is similar in form. A transverse section of the leaf (f) reveals convergence with the leaves of *Ranunculus fluitans*. They are reinforced by strands of xylem, have aerenchyma, and have photosynthetic apparatus concentrated in the epidermis.

rhizomes which help stabilize the soil and which produce two types of root: thick corky anchorage roots; and fine short-lived absorption roots.

The vegetation of salt marshes can be quite variable, but whichever plants dominate, they all help trap sediment. Just like mangrove swamps, therefore they gradually raise the level of the mud, and so reclaim land from the sea. They also produce a highly productive ecosystem which is colonized by molluscs, crustaceans and other invertebrates which in turn are eaten by a wide range of wading birds. It is undoubtedly the diversity of these birds which gives salt marshes such a high public profile and makes them the most widely conserved habitats in temperate regions.

11.3.4 Subtidal plants

Subtidal areas of the sea provide perhaps the toughest of all habitats for vascular plants, which must simultaneously overcome the problems of total submersion and high salinity. Even this habitat, though, has been conquered by several families of the mono-

cots, and the **seagrasses** such as *Zostera marina* and *Thallassia testudinum* show perhaps the most perfect vegetative and reproductive adaptations to underwater life.

Morphological adaptations

Seagrasses (Fig. 11.9) are anchored in the mud by permanent rhizomes and thick, fibrous roots. They produce short-lived leaves (Fig. 11.9b, f) which are in many ways similar to those of submerged aquatic plants: they are ribbon-like in shape; they are reinforced by isolated strands of simple xylem; they contain aerenchyma which transports oxygen down to the roots; and their photosynthetic apparatus is concentrated in the epidermis. The main difference is that the epidermis contains salt-excreting cells rather than salt-absorbing cells.

Pollination

The process of water pollination has been more nearly perfected by seagrasses than by any of the aquatic species we have examined. Most species of seagrass are **dioecious**, having separate male and female plants. In the male flowers, the stamens extend well above the small petals and release elongated pollen grains which are joined together in long gelatinous strands (Fig. 11.9d). These are moved by tidal currents and are ideally formed to get tangled up in neighbouring plants and so contact the extended stigmas of female flowers.

Ecological role of seagrasses

Like the other marine angiosperms we have examined, seagrasses play an important role in stabilizing shorelines and so helping to reclaim new land from the sea. They form productive forests in estuaries and shallow coasts from the tropics right up to the Arctic. Seagrasses epitomize the success of the angiosperms which, as we have seen in the last part of this book, have proved adaptable enough to evolve to dominate almost any habitat. They have been able to recolonize the habitat which the vascular plants originally left over 400 million years ago—the sea. As we have seen, however, although angiosperms are in most areas the dominant types of plants, nowhere have they been able to totally exclude the other taxa of plants, which are themselves continuing to evolve. Even in these seagrass communities, the original inhabitants of the shore, the seaweeds, are common.

They have taken advantage of the firm foundation seagrasses provide and grow as epiphytes on their leaves!

11.4 POINTS FOR DISCUSSION

1 Why do you think angiosperms and ferns have been so successful in adapting to life in water, while conifers remain stranded on land?
2 There are many species of water mosses. Where do you think they are most often found and what do they look like?
3 Why do you think there are no marine ferns?
4 Do you think vascular plants will ever outcompete seaweeds on rocky shores?

FURTHER READING

Cook, D.K. (1990) *Aquatic Plant Book.* SPB Publishing, The Hague.
Crawford, R.M.M. (1989) *Studies in Plant Survival.* Blackwell Scientific Publications, Oxford.
Dawes, C.J. (1981) *Marine Botany.* John Wiley, New York.
Hogarth, P. (2000) *The Biology of Mangroves.* Oxford University Press, Oxford.
Sculthorpe, C.D. (1967) *The Biology of Aquatic Vascular Plants.* Edward Arnold, London.

A classification of plants

As we saw in Chapter 2, there are many different ways to classify plants. The following scheme is based on the principles outlined in Chapter 2, and encompasses only the groups of plants covered in this book, so does not attempt to be complete. Only groups that include living plants are listed, together with some alternative names and a brief description of characteristics of plants in the group.

Two of the three domains of organisms include photosynthetic organisms considered herein to be plants, these are Bacteria and Eukarya (prokaryotes and eukaryotes, respectively).

DOMAIN: BACTERIA

Kingdom: Cyanobacteria (Cyanophyta; Myxophyta; blue-greens)

Photosynthetic bacteria. Photosynthesis is based on chlorophyll *a* and phycobilins. Some species can fix atmospheric nitrogen. Possess peptidoglycan cell walls. Most species are unicellular or filamentous. There is no sexual reproduction and none has flagella. Many abundant and ecologically important species (e.g. Fig. A.1.).

Kingdom: Prochlorophyta

Photosynthetic bacteria. Photosynthesis is based on chlorophylls *a* and *b*, and carotenoids. Very few species.

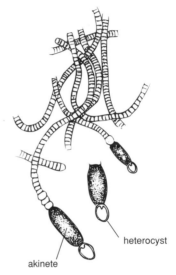

Fig. A.1 *Cylindrospernum*, a cyanobacterium capable of fixing atmospheric nitrogen in their terminal heterocysts.

DOMAIN: EUKARYA

Kingdom: Protista

These are eukaryotic unicellular or multicellular organisms, excluding the plants, fungi and animals which undergo sexual reproduction. The phyla are listed below; many of which move by means of flagella.

• *Phylum: Pyrrophyta* (*Dinophyta; dinoflagellates* (see Fig. A.2)). Photosynthetic members of this group have chlorophylls *a* and *c*. They are mostly unicells with two flagella, one of which encircles the cell in a groove. The cell covering is cellulose, and they have a unique mode of cell division. There are

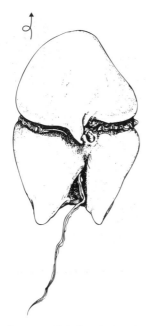

Fig. A.2 A surface view of a *Glenodinium sanguineum*, an organism responsible for red lakes.

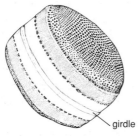

Fig. A.3 A centric diatom.

2000–4000 species, some of which are toxic. Others, the zooxanthellae, form mutual associations with marine animals such as corals. Together with the next four groups these organisms are the most numerous and ecologically important members of a loose assemblage of organisms often described as algae.

• *Phylum: Bacillariophyta (diatoms)*. Photosynthesis is based on chlorophylls *a* and *c*. They are unicellular or colonial organisms, many of which form chains and more elaborate assemblages (Fig. A.3). They possess rigid two-part cell walls based on silica, but no flagella except in some male gametes. There are over 100 000 species, some of which are toxic, others forming mutualistic associations.

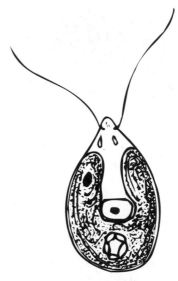

Fig. A.4 A stylized representation of asexual reproduction in *Chlamydomonas*.

• *Phylum: Chlorophyta (green algae)*. Photosynthesis is based on chlorophylls *a* and *b*. They have polysaccharide (usually cellulose) cell walls. They store carbohydrates as starch inside their plastids. They are mostly free-living. There is a vast range of forms, including flagellate and non-motile unicells, colonies, and coenocytic, filamentous, plate-like and parenchymatous organisms (Fig. A.4). There are about 17 000 known species.

• *Phylum: Rhodophyta (red algae)*. Photosynthesis is based on chlorophyll *a* and phycobilins. They have cell walls composed of cellulose or pectins, often with a calcium carbonate layer providing a tough 'exoskeleton' (Fig. A.5). There are unicells and multicellular organisms, the latter often comprising closely packed filaments bound into a pseudo-parenchymatous three-dimensional structure. There are no motile flagel-late cells. There are between 4000 and 6000 known species.

• *Phylum: Phaeophyta (brown algae)*. Photosynthesis is based on chlorophylls *a* and *c*. They have cell walls composed of a cellulose matrix containing algin. Most species are macroscopic multicellular organisms, many parenchymatous with cell differentiation (Fig. A.6). There are only around 1500 species but several are very large and successful.

Fig. A.5 *Chondrus crispus*; the walls have a soft mucilaginous matrix conferring flexibility.

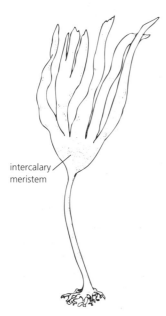

Fig. A.6 The kelps are the marine equivalent of trees. *L. digitata* is a medium-sized organism found subtidally around the coast of Britain.

Kingdom: Plantae

Plants are multicellular organisms with parenchymatous construction and advanced differentiation of cells and tissues. All have photosynthesis based on chlorophyll *a* and *b*, predominantly cellulose cell walls and store carbohydrates as starch inside their plastids, like their closest relatives, the green algae. They have an alternation of generations in their life cycle. Most are terrestrial.

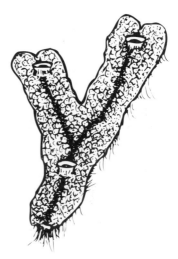

Fig. A.7 The thalloid liverwort *Marchantia polymorpha*.

BRYOPHYTES

All have a sporophyte generation partially or wholly dependent upon and attached to the more dominant gametophyte generation. Their phyla are listed below; they have motile male gametes, no vascular tissue in the gametophyte, and no roots.
- *Phylum: Hepatophyta (liverworts)*. Gametophytes have two forms, leafy or thalloid. The sporophytes are tiny, lack stomata or vascular tissue, but possess elaters. Rhizoids are single-celled. There are around 6000 species, mostly restricted to damp environments (Fig. A.7).
- *Phylum: Anthocerophyta (hornworts)*. Gametophytes are thalloid. The sporophytes possess stomata but no vascular tissue. There are around 100 species, restricted, like liverworts, to damp environments.
- *Phylum: Bryophyta (Musci; mosses)*. Gametophytes are leafy. The sporophytes have stomata, vascular tissue and many have complex and varied modes of dehiscence (Fig. A.8). Vascular tissue is also present in some gametophytes. There are around 10 000 species, many of which are ecologically important.

VASCULAR PLANTS

All vascular plants have true xylem and phloem. The sporophyte is usually the dominant generation, usually possessing cuticle and stomata in the photosynthetic areas.

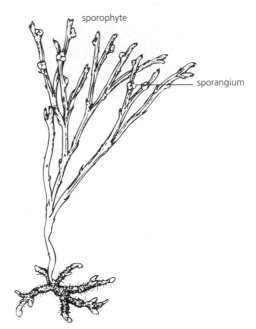

Fig. A.9 The diploid sporophyte of the whisk fern *Psilotum*.

Fig. A.8 The topmost part of a gametophyte—the living material at the surface of the bog—of the bog moss *Sphagnum papillosum*.

Spore-producing plants

Vascular plants without seeds, instead dispersing via spores (see the phyla listed below).

• *Phylum: Psilotophyta*. A few species of homosporous vascular plants with a simple sporophyte having no roots and no leaves (Fig. A.9). Subterranean gametophytes.

• *Phylum: Lycophyta (Microphyllophyta, lycopods— includes club mosses* (Fig. A.10), *quillworts, spike mosses)*. Homo- and heterosporous organisms, all with lateral spor-angia. All have microphylls and roots. Some have superficial photosynthetic gametophytes, others have subterranean ones. There are about 1000 living species, all of which are herbaceous, although many fossil forms were arborescent.

• *Phylum: Sphenophyta (Arthrophyta, horsetails)*. The living species are all homosporous, and eustelic, with reduced leaves and photosynthetic stems (Fig. A.11).

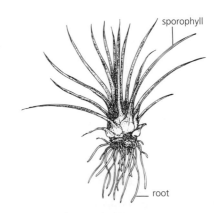

Fig. A. 10 A side view, with one sporophyll enlarged of the quillwort *Isoetes* to show the air channels that run through the leaves and an outline of the sporangium.

The terminal sporangia are reflexed on sporangiophores organized into cones. There are 15 living species, all herbaceous, although many fossil forms were arborescent.

• *Phylum: Pterophyta (Filicophyta, ferns* (Fig. A.12) *)*. Mostly homosporous organisms, although a few are hetero-sporous. All have megaphylls and most have super-ficial photosynthetic gametophytes. There are around 11 000 species, most of them terrestrial.

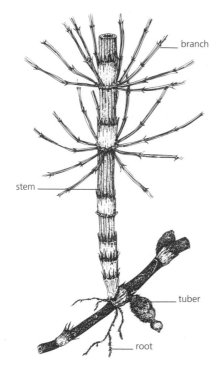

Fig. A. 11 The horsetail *Equisetum*.

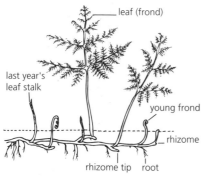

Fig. A. 12 Bracken (*Pteridium aquilinum*) showing the fern's underground rhizome, bearing roots, the remains of last-season's leaves, buds, unfurling leaves and fully expanded leaves.

Seed plants

Plants that retain their ovules within the parent plant protected by two layers of integument, to form a seed.

Gymnosperms. Plants with wholly or partly exposed seeds (see the phyla listed below).

Fig. A.13 The cycad *Dioon*, one of the smaller cycad genera.

Fig. A. 14 A leaf, showing the two lobes, of *Ginkgo biloba*.

• *Phylum: Cycadophyta (cycads).* Plants with pinnately compound palm-like or fern-like foliage and upright stems with little secondary xylem (Fig. A.13). Ovules are exposed and reproductive parts of all but one species are organized into large cones. Male gametes are motile. There are around 140 species, all terrestrial.

• *Phylum: Ginkgophyta (maidenhair trees).* Trees with fan-shaped leaves and abundant secondary xylem. Male gametes are motile. Only one living species (Fig. A.14).

• *Phylum: Coniferophyta (conifers).* Trees with simple, usually linear leaves, many evergreen and abundant secondary xylem. The pollen germinates to produce a pollen tube which grows into the ovule. Around 550 terrestrial species which dominate vegetation in some seasonal areas (Fig. A.15).

Fig. A. 16 A typical 'primitive' angiosperm flower. that of the buttercup *Rannunculus acris.*

Angiosperms. Plants with ovules enclosed in a carpel (Fig. A.16).
• *Phylum: Anthophyta (flowering plants).* Plants with xylem containing vessels. They possess highly reduced gametophytes, double fertilization, and seeds borne within fruits. This is the most successful group of plants with around 235 000 species and a vast range of form. The majority of plants are terrestrial, although some are aquatic.

Fig. A. 15 A short shoot of *Pinus sylvestri*, bearing two needle like leaves.

Glossary

Abscission layer Layer of cells at the base of a deciduous leaf or other organ which is weakened to allow the organ to drop off at a predetermined time.

Achene A simple indehiscent fruit in which each seed is surrounded by a relatively unmodified carpel.

Adaptation A feature of an organism which fits it to its environment and hence improves its chances of surviving and producing offspring.

Adventitious Arising from an unusual place, such as roots growing from stems.

Aerenchyma Tissue containing air spaces which helps ensure aeration of water plants.

Agglutinins Glycoproteins on the flagella of *Chlamydomonas* which cause sexually competent individuals to stick to one another.

Akinetes Specialized spores in cyanobacteria which are large and surrounded by a thick wall.

Albuminous cell A parenchyma cell associated with a sieve-tube element in gymnosperms.

Alga (pl. algae) Old-fashioned term for a member of a series of unrelated groups of freshwater and marine non-vascular photosynthetic organisms.

Allelopathy The chemical suppression by one plant of other, potentially competitive, plant.

Allopatric speciation Evolution of a new species that results from the geographical separation of a population of organisms followed by their subsequent evolution.

Allopolyploidy Production of a polyploid by hybridization of two species followed by a doubling of the chromosome set.

Alternation of generations The life cycle that results from sporic meiosis, in which the organism spends part of its time as a haploid individual and part as a diploid individual.

Amphibious Able to live both in water and on land.

Analogy A feature found in two or more species which has the same present function although not necessarily the same evolutionary origin. Compare with homology.

Anomalous secondary thickening Unusual form of secondary thickening, common in lianas, which produces isolated segments of xylem.

Annual Herb that lives for a single growing season, surviving drought or cold seasons as a seed.

Antheridiogen Chemicals produced by some fern gametophytes that cause spores to germinate and gametophytes to produce antheridia.

Antheridium (pl. antheridia) A male sexual organ of a bryophyte or a pteridophyte.

Apomixis A type of reproduction in plants in which seeds and fruit are formed without fertilization taking place.

Arborescent Having a permanent tree-like structure.

Archaebacteria Prokaryotic organisms other than the Eubacteria. Some are assumed to be the closest living relatives of the Eukaryotes.

Archegonium (pl. archegonia) A female sexual organ of a bryophyte, pteridophyte or gymnosperm.

Aril An outgrowth of the seed coat used by some gymnosperms and angiosperms to promote dispersal.

Artificial selection The selection for desirable characters imposed by breeders on domesticated plants and animals.

Asexual reproduction Any reproductive process that involves no meiosis or fusion, only mitosis. Offspring are similar to the parent. Compare with sexual reproduction.

Autopolyploidy Production of a polyploid by a doubling of the chromosome set of a single species.

Autotroph An organism that synthesizes the organic molecules it needs from the inorganic molecules in its environment.

Basal meristem A growing point situated low down the shoot of plants such as grasses for protection.

Bacteriorhodopsin A pigment like our own visual pigment used by some Archaebacteria to trap light energy.

Benthic Descriptive of organisms that are attached to the floor of the sea or stretches of fresh water.

Berry Indehiscent fruit containing no hard parts apart from the seed.

Biennial Herb that lives for 2 years, building up an underground energy store after its first growing season and flowering and seeding in its second.

Bloom Huge build-up in the numbers of planktonic organisms that can cause environmental hazards.

Boreal forest Evergreen forest found in cool temperate regions, also known as Taiga.

Bryophyte Member of the group of non-vascular land plants with a dominant gametophyte generation which includes the liverworts, hornworts and mosses.

Bulb An underground stem with a food store with roots beneath and fleshy leaves above.

Bulbil Replica of a parent plant which grows out of its surface and separates from the parent plant in vegetative asexual reproduction.

Bundle sheath cells Layer of cells around a vascular bundle where C_4 plants concentrate their photosynthesis.

C_3 photosynthesis The commonest form of photosynthesis found in most plants of cool temperate regions.

C_4 Photosynthesis Form of photosynthesis in which gas exchange and carbon fixation are spatially separated. Common in plants of hot, dry regions.

Calyx The sepals of a flower.

Cambium A meristem that produces parallel rows of cells.

Canopy The main upper layer of leaves and branches in a forest.

Capitulum Inflorescence like that of daisies in which all the flowers are clustered on a single platform.

Capsule A spore-containing structure of a bryophyte, or a seed-containing structure of an angiosperm.

Caretenoids A class of accessory photosynthetic pigments found in chloroplasts. Includes carotenes (yellow and orange) and xanthophylls (yellow).

Carnivory The feeding on animals by plants or other animals.

Carpel A female part of a flower, which contains the ovules.

Cast fossils The internal casts in rock of the hollow parts of plants.

Catkin Male inflorescence of a tree or shrub that releases wind-borne pollen.

Cauliflory Having flower stalks growing directly out of the trunk. Seen in many rainforest trees.

Central cell Cell in the embryo sac which contains the polar nuclei.

Centric Radially symmetrical.

Chasmoendolithic Living within cracks in rock.

Chlorophyll The green pigment found in plant cells which is the main receptor of light energy in photosynthesis.

Chromosome A thread of DNA which appears as a rod during cell division, that contains hereditary information, or genes.

Cladistics A method of classifying organisms using shared derived features to work out an accurate estimate of their ancestry.

Cladogram A branching diagram that shows the phylogenetic relationships that have been worked out by cladistics.

Climax tree Tree that is adapted to outcompete pioneers and replaces them in the later stages of secondary succession.

Climber A plant that obtains its support from another self-supporting plant, but that is not otherwise dependent on it.

Coelenterate Member of a group of animals that contains jellyfish, hydra and the animal part of corals.

Coenocytic Having a body made up of a single, multinucleated cell, as the nuclei are not separated by walls or septa.

Coevolution The simultaneous evolution of two or more species of organisms which results from their close biotic interactions.

Companion cell A cell associated with a sieve-tube element in the phloem of angiosperms, and derived from the same mother cell.

Competition An interaction between two or more organisms which require the same, usually limited, resource.

Compression fossil The hardened and flattened remains of an organism.

Conceptacle A protective cup that protects the eggs of some brown seaweeds until they are fertilized.

Continuous character A characteristic of an organism which is influenced by a large number of genes.

Convergence The independent evolution of functionally similar structures in distantly related organisms as a result of similar selection pressures. Compare with parallel evolution.

Corm A short, thickened, underground stem base specialized for food storage.

Corolla The petals of a flower.

Corticated filament Filament surrounded by a further layer or layers of cells to make it three-dimensional.

Cotyledon The first leaf of a plant embryo which often contains food stores.

Crassulacean acid metabolism (CAM) Form of photosynthesis in which gas exchange and carbon fixation are temporally separated. Common in desert succulents and epiphytes.

Crown group radiation The major groups of the eukaryotes that split apart from each other about 1 billion years ago.

Cryptoendolithic Living within rock.

Cuticle Waxy outer covering over the epidermis of a land plant which reduces water loss.

Cyanobacteria Photosynthetic prokaryotes that possess chlorophyll and that perform conventional photosynthesis. Sometimes called blue–green algae.

Cyme An inflorescence in which the terminal flowers open first and the basal ones last.

Deciduous Condition in which leaves are all shed at a particular season. Compare with evergreen.

Deciduous forest Forest found in temperate regions in which most plants lose their leaves over winter.

Defence An adaptation by a plant which reduces the exploitation of its photosynthesis by herbivores or parasites.

Deoxyribonucleic acid (DNA) The molecule of inheritance for plants and all other organisms. DNA is present in the chromosomes, and is composed of a double chain of nucleotides whose sequences code genetic information.

Diatom One of a group of unicellular eukaryotes with silicaceous walls. Includes important members of the plankton.

Dinoflagellate A member of a group of unicellular eukaryotes, most of which have two flagella, cellulose cell walls and permanently condensed chromosomes. Includes important members of plankton and the photosynthetic parts of many corals.

Dioecious Having male and female reproductive part on separate individuals.

Diploid State in which each cell of an organism contains two copies of each chromosome. Compare with haploid.

Discontinuity Gap in time in a fossil-containing bed, during which no new sediment is laid down.

Discrete character A characteristic of an organism that is influenced by a single gene.

Disease The effects of the exploitation of plants' photosynthesis by microbes such as fungi, bacteria and viruses.

Divergence The evolution of dissimilar form by species or groups of species with common ancestry.

Division of labour Specialization of cells in multicellular organisms to perform separate tasks.

DNA hybridization A technique used to determine the relatedness of two organisms by comparing the melting point of hybrid DNA strands with that of each organism.

Double fertilization The two simultaneous fusions of sperm nuclei with ovule nuclei in angiosperms. The fusion of egg and sperm forms the diploid embryo, while that with the polar nuclei forms the triploid endosperm.

Drip tip Extended end to the leaf of a rainforest plant, which is thought to help it shed rainwater.

Drop root Aerial root which emerges from the lateral branch of a mangrove.

Drupe Fleshy fruit with a hard inner coat which protects one or a few seeds.

Dry season Period when tropical and subtropical areas receive little rainfall.

Ectohydric Transporting water up the outside.

Egg-beater trichomes Waxy water-repellent hairs which give buoyancy to some floating water ferns.

Elaiosome A small food body attached to a seed or fruit which attracts ants.

Elaters Specialized cells in the capsule of a liverwort, which act to catapult the spores into dry air.

Embolism Air bubble that can block the water movement through xylem.

Embryo A young diploid sporophyte plant before it grows or germinates.

Emergent Tree that extends above the main canopy of a forest.

Encrusting Forming a two-dimensional plate of cells on the surface.

Endemic A species found only on a single island or in a single region which has usually been formed by allopatric speciation.

Endohydric Transporting water up the inside.

Endophagy The feeding on plants from the inside, usually by microscopic animals.

Endosperm The triploid tissue, containing stored food, which is used as a food source by the developing angiosperm embryo.

Endospory The condition in which gametophytes grow inside their spore coat and live off stored food.

Endosymbiosis The process by which a larger organism engulfs smaller organisms to form a single compound organism. It is thought eukaryotic cells were formed by endosymbiosis of prokaryotes into cells to form mitochondria and chloroplasts.

Epiphyte A plant that grows on another plant but that is not dependent on it for organic carbon or water.

Eubacteria The main group of living Prokaryotes, which includes the cyanobacteria and most heterotrophic bacteria.

Eudicotyledon One of the two major groups of angiosperms, in which each seed has two seed leaves or cotyledons.

Eukaryote An organism whose cells have a nucleus and organelles which are surrounded by membranes and rod-shaped chromosomes. Compare with prokaryote.

Eustely A condition in which the primary vascular tissue is arranged in discrete strands around a pith. Found in seed plants.

Evergreen Condition in which a plant sheds its leaves sequentially over a long time period, so that some leaves are always present. Compare with deciduous.

Evolution Cumulative change in populations of organisms over multiple generations. This has resulted in the production of the diversity of life today.

Evolutionary arms race The result of coevolution between two interacting species. It can produce features which appear unusual and even non-adaptive.

Evolutionary taxonomy A method of classifying organisms based on a combination of ancestry and overall similarity.

Extinction The elimination of a species from earth when the last individual dies.

Fibre A very long, narrow lignified cell which has a role only in support. Compare with vessel element, tracheid and fibre.

Filament Single file of cells, which may be branched or unbranched.

Fixation The increase in frequency of an allele until it is the only form of a gene found.

Flagellum (pl. flagella) Long thread-like organelle which protrudes from the surface of a cell. The flagella of eukaryotes are capable of wave-like motion and act to propel many spores, gametes or small organisms through liquids.

Fossil The preserved remains or traces of an ancient organism, usually found in rock.

Frond Flat leaf-like region of seaweeds and ferns, which is specialized for photosynthesis.

Frost heave The levering up of plants by a succession of freeze–thaw cycles in polar regions.

Fruit A mature ripened ovary or collection of ovaries which provides protection and/or dispersal for the seeds within.

Frustule Outer wall of a diatom.

Fusion The process in sexual reproduction in which two haploid cells or gametes meet and their nuclei fuse to form a new diploid cell or zygote.

Gametangium The sexual organ of a plant which generates gametes.

Gamete A haploid reproductive cell—an egg or a sperm— which in sexual reproduction fuses with another to form a diploid zygote.

Gametic meiosis The division of a cell from a diploid individual to produce haploid gametes. These then fuse to form diploid zygotes that divide by mitosis to form diploid individuals. Eventually these produce more gametes. Organisms with gametic meiosis spend most of their time in the diploid state.

Gametophyte The gamete-producing stage in the life cycle of an organism which shows sporic meiosis and hence has alternation of generations.

Gas vesicle Cylindrical structure, acting to help flotation, which is found in large numbers in the gas vacuoles of some cyanobacteria and halobacteria.

Gemma (pl. gemmae) A specialized outgrowth of a non-flowering plant which can break off and grow into a new plant, so resulting in asexual reproduction.

Gene A discrete unit of hereditary information, consisting of a sequence of DNA nucleotides that usually codes for a polypeptide molecule.

Genotype The genetic make-up of an organism.

Halophyte A plant that can survive in very saline water.

Haploid State in which each cell of an organism contains only a single copy of each chromosome. Compare with diploid.

Haustorium A structure of a parasite that penetrates host tissue and extracts material from it.

Heathland Nutrient-poor temperate habitats containing scattered trees and perennial vegetation.

Hedgehog plant Short, spiky, cushion-shaped plants characteristic of Mediterranean regions.

Hemiparasite Plant that obtains water and nutrients from its host but that is still able to photosynthesize. Compare with holoparasite.

Herb A non-woody seed plant with a relatively short-lived shoot system.

Herbivory The feeding on plants by (usually macroscopic) animals.

Heterocyst A cell specialized for nitrogen fixation in some filamentous cyanobacteria.

Heteromorphic A term descriptive of a life cycle in which the haploid and diploid stages are dissimilar in form.

Heterophylly The production of leaves of differing morphology.

Heterospory Production of two types of spores, megaspores and microspores, which develop into unisexual gametophytes.

Heterotroph An organism that cannot manufacture the organic molecules it needs, and so must obtain them from other organisms, either live or dead.

Hexaploid A plant containing six copies of each homologous chromosome.

Holdfast A basal structure of a multicellular alga which attaches it to a solid substrate.

Holoparasite Plant that obtains all of its nutrition from its host. Compare with hemiparasite.

Homology A feature found in two or more species which has the same evolutionary origin but not necessarily the same present function. Compare with analogy.

Homospory Production of a single type of spore which develops into a bisexual gametophyte.

Hopeful monsters Organisms that have resulted from large mutations. They are unlikely to survive but could give rise to new groups of organisms.

Hormogonium A portion of a cyanobacterial filament that becomes detached and grows into a new filament.

Hyaline cells The outermost thickened cells of some non-flowering plants which are used for external water conduction.

Hybridization The production of offspring following mating between two different species of organisms. Hybrids are usually sterile unless polyploidy occurs.

Hydroid Water-conducting tissue of mosses, analogous to the xylem of vascular plants.

Hydropoten cell Absorptive hair on the underside of a floating leaf which takes up ions from the water.

Impression fossil The imprint in rock of an organism which may have decayed away.

Indusium (pl. indusia) Outgrowth of a fern frond that protects the sporangia.

Inflorescence A cluster of flowers on a flower stalk.

Integument The outermost layers of tissue which envelop and protect the ovule, and later develop into the seed coat.

Intercalary meristem Growing point at the node of a plant that is responsible for elongation.

Isidium (pl. isidia) Finger-like propagules of lichens.

Isomorphic A term descriptive of a life cycle in which the haploid and diploid stages are identical in form.

Key adaptation Evolutionary novelty peculiar to a particular group of plants which may have been significant in its survival and subsequent radiation.

Leaf A lateral appendage of the stem of a vascular plant, usually flattened, which is specialized for photosynthesis.

Legume Two-valved fruit of members of the pea family.

Leptoid Sugar-conducting tissue of mosses analogous to the phloem of vascular plants.

Liana A woody climber common in rainforests.

Lichenometry A method used to date ancient stones using the growth of lichens.

Life cycle The whole sequence of stages in the growth and development of an organism from zygote formation to gamete formation.

Lignin Plant polymer that strengthens the walls of some plant cells, particularly woody xylem cells.

Living fossil A member of a once important group of organisms which has changed little since most of its relatives became extinct.

Long branch attraction The problem in molecular taxonomy that distantly related groups with no intermediate forms appear more closely related.

Macroevolution Large-scale evolutionary change, involving major trends and occurring over a geological time scale.

Mangrove Tree that grows in the intertidal zone of sandy or muddy tropical shores.

Mangrove swamp Vegetation of a muddy tropical shore that is dominated by flooding-tolerant trees.

Maquis The shrubby vegetation of dry Mediterranean regions.

Mass extinction An event during which a large number of species simultaneously died out. Now thought to be caused by physical processes such as meteorite strike or volcanic activity.

Meadow Species-rich grasslands of temperate regions kept free of trees by grazing pressure.

Mediterranean Region on western shores of the lower temperate areas which have a mild wet winter and hot dry summer.

Megaphyll A type of leaf found in ferns, horsetails and seed plants that contains multiple vascular strands. Compare with microphyll.

Megaspore A spore produced by a heterosporous plant that develops into a female gametophyte. Includes the ovule of seed plants.

Meiosis The process by which a diploid cell undergoes two successive divisions to produce four haploid cells containing novel chromosomes.

Meristem The undifferentiated tissue, usually at the tips of the plant from which new cells arise.

Meristoderm Cambial layer, like that in trees, below the epidermis of some brown algae which produces secondary thickening.

Microbial mat A layer of microscopic organisms on a surface.

Microevolution Small-scale evolutionary change which results from changes in the frequencies of genes over a small number of generations.

Microphyll A leaf that contains just a single vascular strand. Compare with megaphyll.

Micropyle The opening in the integuments of seed plants through which the pollen tube enters.

Microspecies Varieties of plants formed by asexual reproduction which are adapted to local conditions.

Microspore A spore produced by a heterosporous plant that develops into a male gametophyte. Includes the pollen grains of seed plants.

Mitochondrion (pl. mitochondria) An organelle found in a eukaryotic cell that is important in producing energy.

Mitochondria are thought to have been derived by endosymbiosis of bacteria.

Mitosis The process by which a cell's nucleus divides to produce two virtually identical daughter nuclei, each with the same number of chromosomes as the parent nucleus. The cell also usually divides.

Molecular clock The assumption that DNA of different species diverges at a constant rate.

Monocolpate Having a single furrow or pore in a pollen grain.

Monocotyledon One of the two major groups of angiosperms, in which each seed has one seed leaf or cotyledon.

Monoecious Having male and female reproductive parts in separate flowers on a single individual.

Monsoon The rainy season in some regions of the tropics and subtropics.

Monsoon forest Forest found in outer tropical and in subtropical regions.

Moorland Upland temperate regions dominated by grasses and small shrubs which are kept free of trees by grazing and burning.

Morphology The form or structure of an organism.

Multicellularity The condition in which organisms are made up of large numbers of cells which have divided but not separated.

Multiple base exchange The problem in molecular taxonomy that multiple changes at a point along a DNA strand cannot be differentiated from a single change.

Mutation A change in the composition or arrangement of the DNA within a chromosome. If it occurs in the reproductive cells this can be passed to future generations and result in evolutionary change.

Mutualism A symbiotic relationship in which both partners benefit.

Mycobiont The fungal partner of a lichen.

Mycorrhiza A mutualistic association between a fungus and a root which helps the plant absorb minerals and water from the soil.

Natural group A group of organisms containing all the descendants of a single common ancestor. They are linked by shared derived characteristics.

Natural selection The process that causes evolution. Organisms with favourable adaptations are more likely to survive and pass on their traits to future generations.

Neotony The delay in morphological development that results in paedomorphosis.

Non-convergence The independent evolution of very different structures in distantly related organisms as a result of similar selection pressures. Compare with convergence.

Nucellus The tissue in the ovule that contains the embryo sac.

Nut Dry, indehiscent, woody fruit.

Ombrotroph A pant that obtains its water and nutrients directly from rainfall and without roots.

Oogonium (pl. oogonia) A female sex organ that contains eggs.

Organism A living creature, either unicellular or multicellular.

Outgroup comparison The use of more distantly related organisms to determine the direction of evolution.

Ovule A megaspore held within a megasporangium of seed plants which will develop into a seed if fertilized.

Paedomorphosis Evolutionary process by which an organism becomes sexually mature at an earlier stage in development. Can occur by neotony or progenesis.

Parallel evolution The independent evolution of functionally similar structures in closely related organisms as a result of similar selection pressures. Compare with convergence.

Parasitism A symbiotic relationship in which one organism is exploited by the other.

Parenchyma Tissue formed by cell division in three dimensions. Compare with pseudoparenchyma. In land plants often refers to unspecialized large-celled tissue.

Parsimony A rule used in cladistics that assumes that the phylogeny which involves the fewest evolutionary steps is the most likely.

Pennate Winged.

Perennating bud Dormant bud that can survive drought or cold weather before opening when good conditions return.

Perennial Herb that lives for more than 2 years, surviving drought or cold using underground buds and storage organs.

Peristome The fringe of teeth around the opening of the capsule of a moss.

Permafrost Ground below the surface of polar regions which remains permanently frozen.

Petal A modified leaf that makes up the next to outermost whorl of a flower. Usually conspicuously coloured to attract pollinators.

Petiole Leaf stalk.

Petrifaction Fossil formed by the inundation of mineral-rich water into living tissue.

Phenetics A method of classifying organisms based on their overall similarity.

Phenolics Defence chemicals that work by precipitating digestive enzymes in the stomach of herbivores.

Phenotype The physical appearance and behaviour of an organism that results from the interaction between its genetic make-up (genotype) and its environment.

Pheromone Signalling molecule that allows recognition at a distance between organisms.

Phloem The sugar-conducting tissue of vascular plants.

Photobiont The photosynthetic partner of a lichen, either a green alga or cyanobacterium.

Photosynthesis The biological process in which light energy is used to produce carbohydrates from carbon dioxide and water.

Phragmoplast A structure arising between two dividing land plant nuclei at which a cell plate dividing the cell into two is formed. Compare with phycoplast.

Phreatophyte Desert plant which exploits deep water resources using a long tap root.

Phycobilins Accessory photosynthetic pigments found in cyanobacteria and red algae.

Phycoplast An ingrowth of the outer walls of dividing green alga cells which separates them into two. Compare with phragmoplast.

Phylogeny The evolutionary history or tree of a group of organisms.

Picoplankton Tiny floating organisms usually a micrometre or less in diameter.

Pinnate leaf Leaf divided into several leaflets on either side of its petiole.

Pioneer tree Tree that is adapted to invade disturbed habitats and thrive in the early stages of secondary succession. Compare with climax tree.

Planation Flattening of branch systems which is thought to have contributed to the evolution of megaphylls (see webbing).

Plankton Free-floating, usually microscopic organisms found in surface waters.

Plasmid A small circular molecule of DNA found in bacteria, usually separate from the chromosome.

Plasmodesmata The minute cytoplasmic threads that connect the protoplasts of adjacent cells through holes in the cell walls.

Pneumatophore Upward extensions of roots used by some trees of swampy habitats to ensure adequate aeration of the root system.

Poikilohydric Describes a plant whose water content varies with that of the surrounding environment.

Polar nuclei Two haploid nuclei in the embryo sac of angiosperms, which fuse with a sperm cell during fertilization to form the triploid endosperm.

Pollen drop A sticky fluid produced by the ovule which traps pollen grains.

Pollen grain A microspore of a seed plant.

Pollen tube A tube formed after the pollen grain germinates, which carries the male gametes to the ovule.

Pollination The transfer of pollen from the male to the female part of seed plants.

Pollinium (pl. pollinia) Pollen bearing organ of an orchid.

Polymorphic Able to develop a different morphology depending on the environmental conditions.

Polyploidy The multiplication of sets of chromosomes to produce more than two pairs in a single nucleus.

Pome Fleshy fruit whose outer coat is derived from the receptacle.

P-protein A protein found in the sieve-tube elements of angiosperms.

Prairie A dry temperate grassland.

Preadaptation A feature of an organism that proves beneficial when used for a different function from that which it first suited.

Primary forest Area of forest that has been undisturbed by humans for a long period of time. Compare with secondary forest.

Primary succession The invasion and alteration of new environments by vegetation.

Prochlorophyta A group of photosynthetic bacteria similar to cyanobacteria, but which more closely resemble the chloroplasts of green plants.

Progenesis The speeding up of sexual development, a process that results in paedomorphosis and that may have been responsible for the evolution of herbaceous angiosperms.

Prokaryote An organism whose cells lack a nucleus and membrane-bound organelles, and have a single circular chromosome. Compare with eukaryote.

Prop root Aerial root that emerges from the trunk of a mangrove.

Protostely Condition in which the vascular tissue is concentrated into a solid central column. Found in the shoots of early vascular plants and in roots.

Pseudoparenchyma Three-dimensional tissue formed by the conglomeration of branched filaments.

Punctuated evolution A model of evolution in which there are long periods with little or no change followed by brief intervals in which major rapid changes occur.

Raceme An inflorescence in which the basal flowers open first and the terminal flowers last.

Raised bogs A bog which receives its water supply solely from precipitation.

Raphe Groove on the frustule of a diatom.

Recapitulation The addition, over evolutionary time, of a novel stage at the end of development.

Receptacle The part of a flower stalk that bears the flower.

Recombination The formation in gametes of new combinations of genes by crossing over and shuffling of chromosomes during meiosis.

Reproductive isolation The barriers that prevent two populations interbreeding. If interbreeding is not possible even when the two populations meet, they will have formed two species.

Respiration The intracellular process in which molecules are oxidized to liberate energy. The complete breakdown of sugars to carbon dioxide and water is called aerobic respiration.

Restriction enzymes Enzymes that cleave the DNA sequence at specific sites.

Rhizine Extension of the ventral surface of a lichen which is used to absorb water and nutrients.

Rhizoid Extension of the ventral surface of a charophyte, bryophyte or primitive vascular plant, which is used to anchor the plant and absorb water and nutrients.

Rhizome A horizontal underground stem.

Rhizophore An anchoring structure in club mosses which bear roots.

Ribonucleic acid (RNA) A single-stranded nucleic acid molecule formed on chromosomal DNA and necessary for protein synthesis.

Root A multicellular structure, normally below ground, which anchors vascular plants and absorbs and conducts water and nutrients.

Salt marsh Intertidal regions of sandy temperate shores, dominated by xerophytic and succulent perennial herbs.

Samara A winged seed.

Saprophyte An organism that obtains its food directly from dead matter.

Savannah vegetation type seen in the dry tropics, with scattered trees and grassland.

Scalariform plate Half open end-wall of a vessel element.

Seagrass Member of one of several species of marine monocotyledons.

Secondary endosymbiosis Capture of organelles by engulfing another eukaryote and retaining its organelles.

Secondary forest Forest which has regrown after the destruction of an area of undisturbed primary forest.

Secondary succession The process that follows the clearing of a small area of vegetation and which results in its ultimate reestablishment.

Secondary thickening An increase in a plant's girth due to new tissues laid down by lateral meristems which are usually located in a ring around the main body axis

Seed A structure formed as the ovule of a seed plant matures after fertilization.

Seed coat The outer layer of a seed, developed from the integuments of the ovule.

Sepal A modified leaf that makes up the outermost whorl of a flower, and which protects the bud.

Sexual reproduction A type of reproduction that involves meiosis and recombination, followed by fusion of gametes at some stage of the life cycle. This increases the variability of offspring. Compare with asexual reproduction.

Shared derived character A character of two or more organisms which alone gives clues about their phylogeny.

Shoot The above-ground part of the plant, consisting of stem and leaves, which has a photosynthetic function.

Shrub A short woody plant, with several stems emerging near the ground.

Simple plate Fully open end-wall of a vessel element.

Siphonostely Condition in which the vascular tissue is concentrated into a hollow cylinder surrounding a pith. Found in the shoots of some spore-producing vascular plants.

Soredium (pl. soredia) An asexual propagule of a lichen containing algal cells and fungal hyphae.

Speciation The process by which new species are produced in nature.

Species A group of organisms with similar characteristics which have the potential to breed only with each other.

Sporangiophore A branch bearing one or more sporangia.

Sporangium (pl. sporangia) A structure in which spores are produced.

Spore A specialized haploid reproductive cell that divides by mitosis to form offspring.

Sporic meiosis The division by meiosis of diploid cells of an individual to produce haploid spores. The spores divide to produce haploid individuals, or gametophytes, which eventually produce gametes. These fuse to form diploid zygotes which divide by mitosis to produce diploid individuals, or sporophytes. Organisms with sporic meiosis have an alternation of generations of haploid and diploid individuals.

Sporophyll A leaf-like structure that bears spores.

Sporophyte The spore-producing stage in the life cycle of an organism which shows sporic meiosis and hence has alternation of generations.

Sporopollenin The tough waterproof substance that makes up the outer wall of the spores and pollen grains of land plants.

Steppe See prairie.

Stigma The region of the carpel where pollen grains are caught and on which they germinate.

Stipe The cylindrical stem-like regions of some seaweeds, which are specialized for support and transport of sugars.

Stolon A horizontal above-ground stem.

Stoma (pl. stomata) A minute pore flanked by guard cells in the epidermis of vascular plants and some bryophyte sporophytes which allows gas exchange.

Strobilus (pl. strobili) A cone-like structure that bears spore-producing sporophylls.

Stromatolite Rocks formed by the fossilization of microbial mats.

Subcanopy Area beneath the canopy of a forest, colonized by short trees and shrubs.

Succulent A plant with fleshy water-storing stem or leaves.

Symbiosis The relationship between two organisms living in close association, which can be harmful to one (parasitism) or advantageous to both (mutualism).

Sympatric speciation Evolution of a new species within the geographical range of its parent species. In plants this usually results from hybridization of two species followed by polyploidy.

Taiga Evergreen forest found in cool temperate regions, also known as boreal forest.

Tannins Defence chemicals that work by precipitating digestive enzymes in the stomach of herbivores.

Taxon A group of related organisms.

Tetrads A group of four spores formed from a single mother cell by meiosis.

Tetraploid A plant containing four copies of each homologous chromosome.

Thallus A flat two-dimensional body of many algae, certain liverworts and a very few aquatic angiosperms.

Thylakoids Sac-like membranous structures found in cyanobacteria and chloroplasts which hold the photosynthetic pigments.

Tracheid An elongated, thick-walled xylem cell, which has a role in both support and water transport. Compare with vessel element.

Tree A large woody plant with a single trunk.

Trichocyst A dinoflagellate organ which may be connected with capture of prey.

Trichome A hair or similar outgrowth of the epidermis.

Tricolpate Having three furrows or pores in a pollen grain.

Triploid A plant containing three copies of each homologous chromosome.

Tropical rainforest A diverse forest habitat of the central tropics, dominated by angiosperm trees.

Tundra A subpolar habitat dominated by short vascular plants and bryophytes.

Umbel A flower head in which the flower stalks all arise from the same point, and produce a flat crown of flowers.

Vascular cambium A lateral meristem that produces secondary xylem (wood) and phloem, located in a ring towards the periphery of the stem branches and roots of a plant.

Vascular plant A member of the group of land plants that contain xylem and phloem and that have a dominant sporophyte. Also known as tracheophytes.

Velamen A multilayered epidermis of the roots of some epiphytes and desert plants which reduces water loss in drought.

Vessel element A wide xylem cell which, when linked end to end with others, produces wide tubes for efficient water conduction in angiosperms.

Webbing The filling-in of the gaps between flattened branches which is thought to have led to the evolution of megaphylls.

Wood Secondary xylem tissue.

Xerophytic Plant adapted to withstand arid environments.

Xylem The water-conducting tissue of vascular plants.

Zooxanthellae Endosymbiotic photosynthetic dinoflagellates found in some coelenterates, including coral animals.

Zygote The diploid cell that results from the fusion of the two gametes in sexual reproduction.

Zygotic meiosis The division by meiosis of a zygote to form four haploid cells. These then divide by mitosis, to produce haploid individuals or a haploid organism which eventually produce gametes. Organisms with zygotic meiosis spend most of their time in the haploid state.

Index

Please note: page numbers in *italics* refer to figures, and those in **bold** to tables. A number prefixed by the letter 'P' refers to the colour plate of that number, e.g. P5 is Plate 5. The colour plate section faces p. 110.

phylogeny 22
  combining results from morphological and molecular techniques 28
  magnolia, rowan, false acacia clover **25**
  and molecular sequence data 27
  plants, reconstruction methods 23–4
    cladistics 23–4
    phenetics 23
physical environment, constraining plant growth 3–4
Picea abies, air-bladder pollen grains *118*
picoplankton, cyanobacteria important members of 38
pine needles
  efficient and weatherproof 149
  a reflection of climatic change 116–17, *117*
pine trees, evolution of samaras 129, *129*
Pinus contorta, use of aerenchyma 171
Pinus silvestris
  air-bladder pollen grains *118*
  desiccation-resistant structure *117*
pioneer trees
  growth of 143
  harbouring ants P8, 143
  seasonal woods 154
pitcher plants, convergent evolution of P8, 19, *20*
planktonic organisms, free-floating 37, *39*
plant evolution
  biological interactions 4–5
  coevolution 5–6
  like a branching net 14
  physical environment 3–4
    limits of physical explanation 4
plant life
  adaptations of 3
  diversity of 3
plant phylogeny
  reconstructions 22–3
    problems using fossils 22–3
Plantago maritima 183
plants
  a classification 186–8Ap
  defined 35
  early 35–6
    bacteria using unconventional photosynthesis 35
    evolution of conventional photosynthesis 35–6
  fossils, reconstruction difficult 23
  freshwater
    with emergent shoot systems 173
    with floating leaves 173–5
    free-floating 175–6
    submerged 176–81
  parasitic 152–3
  readaptation to life in the water 171
  spore-producing
    adaptive radiation of 90, 92
    sexual reproduction of 101–9
  successful in areas with short growing seasons 155–8
  survival in extreme climates 161
plasmids 7
plasmodesmata, give strength to brown algae 61
plate ends, angiosperms, scalariform or simple *119*
Platycerium see stag's horn fern

Pleopteris polypodioides (resurrection fern), a drought avoider P1, 162–3
pneumatophores 171, *172*, 182
Poaceae *see* grasses
Podastemacae, creeping photosynthetic thallus 181
poisons, novel, and angiosperm evolution 157
polar nuclei 120
polar regions 146, 161
  life in 168–9
    algae 169
    guerilla lycopods 168
    lichens 168–9
    mosses 168
pollen drops 118
pollen grains 111
  conifer 117
  efficiency of transfer improved 125
  monocolpate or tricolpate 122
pollen tubes
  angiosperm 122
  cycads 114
pollination *5*
  angiosperms
    early 125
    efficient 135
  animal, may promote speciation 144
  difficult in fast-flowing water 181
  difficult for submerged plants 177
  and flower shape 126
  modern cycads 125
  submerged still water plants 179–80
  true water pollination 179–80
  water pollination 184–5
pollinia, orchids 127
polymorphism 181
Polysiphonia 66
Polytrichum commune
  adaptations to water transport and retention *77*
  multilayered leaves and differentiated stem tissues 76
pomes 130
population expansion, and rainforest destruction 145
Porphyra, edible fronds 59
Postelsia, sea palm, very tree-like 64
Potamogeton, thin leaves of 175
prairies *147*, 156, 158
preadaptations, evolutionary change, through change in organ function 12, *13*
primary rainforest
  climbing plants 137–9
  epiphytes
    adaptations of 139, *141*, *142*, *143*, *149*
    epiphytic orchids 139–40
    selection pressure on 139
  little will be left outside reserves 144
  rainforest parasites 140–1
  rainforest trees 135–7
Prochlorophyta 36
  chlorophylls *a* and *b* and carotenoids 42
progenesis, accelerated sexual deelopment 17–18
progymnosperms
  Archaeopteris 111, *112*
  rapid secondary growth 115